ENVIRONMENTAL REMEDIATION TECHNOLOGIES, REGULATIONS AND SAFETY

NATIONAL WINDSTORM IMPACT REDUCTION PROGRAM

PLANS AND PROGRESS

ENVIRONMENTAL REMEDIATION TECHNOLOGIES, REGULATIONS AND SAFETY

Additional books in this series can be found on Nova's website under the Series tab.

Additional e-books in this series can be found on Nova's website under the e-book tab.

ENVIRONMENTAL REMEDIATION TECHNOLOGIES, REGULATIONS AND SAFETY

NATIONAL WINDSTORM IMPACT REDUCTION PROGRAM

PLANS AND PROGRESS

RALPH SIMMONS
EDITOR

nova publishers
New York

Library of Congress Cataloging-in-Publication Data

ISBN: 978-1-63484-291-4

Published by Nova Science Publishers, Inc. † New York

CONTENTS

PREFACE

Windstorms such as hurricanes, tornadoes, and severe and fast-moving thunderstorm complexes are among the most destructive and economically damaging hazards that affect the United States on an annual basis. Each year, windstorms claim lives, cause injuries, and result in billions of dollars in property damage. In recent decades, rapid development and population growth in areas at high risk of windstorm hazards have increased the Nation's vulnerability. In 2004, as part of a Government-wide effort to counteract that trend, Congress called for the establishment a National Windstorm Impact Reduction Program (NWIRP), a coordinated approach to achieve major reductions in windstorm-related losses of life and property. In 2006, the SDR Windstorm Working Group delivered to Congress an initial Windstorm Impact Reduction Implementation Plan, which identified research activities needed to improve windstorm hazard mitigation and served as a guide for Federal agencies to coordinate new and existing research to fill knowledge gaps in understanding and forecasting windstorms. This book includes a new report for 2013-2014, which continues to fulfill the NSTC's congressionally mandated responsibility to provide biennial reports on the NWIRP. Overall, it finds that the Program has successfully facilitated improvements in windstorm forecast models, warning systems, evacuation planning, structural-design technology, and community preparedness. This book discusses the Windstorm Impact Reduction Implementation Plan.

In: National Windstorm Impact Reduction Program
Editor: Ralph Simmons
ISBN: 978-1-63484-291-4

Chapter 1

NATIONAL WINDSTORM IMPACT REDUCTION PROGRAM: BIENNIAL REPORT TO CONGRESS FOR FISCAL YEARS 2013 AND 2014*

Subcommittee on Disaster Reduction

EXECUTIVE OFFICE OF THE PRESIDENT NATIONAL SCIENCE AND TECHNOLOGY COUNCIL WASHINGTON, D.C. 20502

Dear Members of Congress,

I am pleased to transmit to you the report, *National Windstorm Impact Reduction Program: Biennial Report to Congress for Fiscal Years 2013 and 2014*, produced by the National Science and Technology Council's (NSTC) Subcommittee on Disaster Reduction (SDR) and its Windstorm Working Group.

Windstorms such as hurricanes, tornadoes, and severe and fast-moving thunderstorm complexes are among the most destructive and economically damaging hazards that affect the United States on an annual basis. Each year, windstorms claim lives, cause injuries, and result in billions of dollars in property damage. In recent decades, rapid development and population growth in areas at high risk of windstorm hazards have increased the Nation's vulnerability. In 2004, as part of a Government-wide effort to counteract that trend, Congress called for the establishment a National Windstorm Impact Reduction Program (NWIRP), a coordinated approach to achieve major reductions in windstorm-related losses of life and property. In 2006, the SDR Windstorm Working Group delivered to Congress an initial *Windstorm Impact Reduction Implementation Plan*, which identified research activities needed to improve windstorm hazard mitigation and served as a guide for Federal agencies to coordinate new and existing research to fill knowledge gaps in understanding and forecasting

* This is an edited, reformatted and augmented version of a report issued by the Office of Science and Technology Policy, of the National Science and Technology Council, August 2015.

windstorms. Subsequently, the NWIRP developed biennial reports for Fiscal Years 2007-2008, 2009-2010, and 2011-2012 detailing examples of relevant Federal activities for reducing windstorm impacts undertaken by the NWIRP member agencies during the given reporting periods.

This new report for 2013-2014 continues to fulfill the NSTC's congressionally mandated responsibility to provide biennial reports on the NWIRP. Overall, it finds that the Program has successfully facilitated improvements in windstorm forecast models, warning systems, evacuation planning, structural-design technology, and community preparedness. These improvements have reduced vulnerabilities to windstorms even as the quantity of people, buildings, and critical infrastructure in the built environment exposed to windstorms across the country has grown dramatically. In coordination with academia and the private sector, the NWIRP agencies look forward to identifying and prioritizing additional research and development needs and achieving even greater successes in reducing windstorm impacts. Sincerely,

John P. Holdren
Assistant to the President for Science and Technology
Director, Office of Science and Technology Policy

1. Background

Recognizing that rapid development and population growth in high-risk areas during the preceding decades had greatly increased the Nation's overall vulnerability to windstorm hazards, in 2004 Congress prescribed the establishment of a coordinated Federal-agency effort– known as the National Windstorm Impact Reduction Program (NWIRP) – to accomplish significant reductions in windstorm-related losses of life and property. To this end, the NWIRP has focused on facilitating improvements in forecast models, warning systems, evacuation planning, structural design, and community preparedness to meet this goal. This report for Fiscal Years 2013-2014 (October 2012-September 2014) continues to fulfill the legislatively mandated responsibility to provide biennial reports to Congress on the NWIRP.

Windstorms pose a grave threat to the safety and security of our citizens and our communities. Improved forecasting of windstorms is essential, but additional observations, models, and tools are also needed to advance the understanding of how the built environment responds to windstorm events and to enable the cost-effective design, construction, and retrofit of windstorm-resistant buildings and infrastructure. Efficient warning systems and social analysis are also needed to improve warning response, mitigate user complacency, better support effective decision making, and reduce the losses associated with devastating windstorms. To help meet these and related issues, the NWIRP agencies, academia, and the private sector continue to build on past activities, including those identified in this report, to identify and prioritize specific measurement-science research and development needs for windstorm impact reduction.

2. WINDSTORMS IN FISCAL YEARS 2013 AND 2014

Windstorms such as hurricanes, tornadoes, and severe thunderstorms are among the most destructive and economically-damaging hazards that affect the United States on an annual basis. Each year, such storms claim lives, cause injuries, and result in billions of dollars in damages to property. Fiscal Years 2013 and 2014 brought windstorms of a historic nature. The most impactful windstorm of the fiscal period of fiscal years 2013-2014 was Hurricane/Post-Tropical Cyclone Sandy, which damaged or destroyed upwards of 700,000 homes in the Northeast after it made landfall in the Northeastern United States in late October 2012. Storm surge created some of the most crippling, cascading impacts in the New York-New Jersey region, including flooding in New York City's subway tunnels, water covering runways at La Guardia and Kennedy airports, and damage to the New Jersey Transit System. Hurricane/Post-Tropical Cyclone Sandy was the worst disaster for mass public transit systems (e.g., bus, commuter rail, subway) in U.S. history. On October 30, 2012, the morning after the storm made landfall, more than half of the Nation's daily transit riders were without service.[1]

Sandy made landfall along the southern New Jersey shore near Brigantine, New Jersey, on October 29, 2012, causing historic devastation and substantial loss of life. In the United States, the storm was associated with 72 direct deaths in eight states: 48 in New York, 12 in New Jersey, 5 in Connecticut, 2 each in Virginia and Pennsylvania, and 1 each in New Hampshire, West Virginia, and Maryland. These deaths make Sandy the deadliest hurricane to hit the U.S. mainland since Hurricane Katrina in 2005, as well as the deadliest hurricane/post-tropical cyclone to hit the U.S. East Coast since Hurricane Agnes in 1972.[2] At the height of the event on October 29-30, 2012, approximately 8.5 million people were without power. Based on National Weather Service (NWS) forecasts and population statistics as the storm approached, it is estimated that approximately 58.7 million people were impacted by wind gusts of at least 58 miles per hour and 19.2 million people were impacted by wind gusts greater than 74 miles per hour[3] – an area so large that tropical-storm force winds extended about 1,000 miles in diameter.

In addition to Sandy, a series of powerful and destructive tornadoes of unprecedented size and intensity struck in and around the Newcastle, Moore, and El Reno, Oklahoma, areas in May 2013, causing many devastating impacts in the central part of the state. Windstorms that struck the United States in calendar year 2013 took 92 lives and injured almost 900 people,[4] causing an estimated $4.5 billion in total direct property losses in the process.[5] The deadly Enhanced Fujita scale 5 (EF-5) tornado that struck the cities of Newcastle and Moore, Oklahoma, on May 20, 2013, had a path length of 17 miles, was 1.3 miles wide, and had peak winds estimated at 210 miles per hour.[6] The tornado killed 24 people, injured at least 230, and destroyed over 1,050 homes – despite the city having a 36-minute warning lead-time as the storm was approaching.[7] The EF-3 tornado that ravaged rural areas surrounding El Reno, Oklahoma, on May 31, 2013, had a path length of 16.2 miles and was 2.6 miles wide, which was the largest tornado width ever recorded.[8] There were seven fatalities associated with this tornado.

Additional windstorms and related severe and fast-moving thunderstorm complexes caused extensive damage in other parts of the United States as well during the fiscal period. A massive storm system spawned a tornado outbreak between April 27-30, 2014, across several Central and Southern U.S. states, impacting thousands of residents in Alabama, Arkansas,

Kansas, Mississippi, Oklahoma, and Tennessee. The small towns of Mayflower and Vilonia, Arkansas, in Faulkner County were hit particularly hard by an EF-4 tornado on the evening of April 27, 2014,[9] which resulted in 12 fatalities. According to the NWS, in Faulkner County the tornado destroyed 261 houses and 121 mobile homes, caused major damage to 77 houses, 29 mobile homes, and 5 apartments, and caused minor damage to 96 houses, 14 mobile homes, and 4 apartments.[10] Overall, during calendar years 2013 and 2014, there were 907 and 888 tornadoes, respectively, that occurred in the United States, 28 of which were killer tornadoes responsible for 102 combined tornado deaths over the two-year span.[11]

3. Program Objectives and Agency Roles

The NWIRP was established by Public Law 108-360, Title II, which is known as the National Windstorm Impact Reduction Act (NWIRA) of 2004. This legislation called for the Director of the Office of Science and Technology Policy (OSTP) to establish an interagency working group, and charged that working group with developing and transmitting to Congress:

1) an implementation plan for achieving the objectives of the program; and
2) a biennial report describing the status of the program and progress achieved.

Following passage of the NWIRA, OSTP established, within the structure of the National Science and Technology Council (NSTC), an interagency Windstorm Working Group of the Committee on Environment, Natural Resources, and Sustainability Subcommittee on Disaster Reduction (SDR), consisting of the National Institute of Standards and Technology (NIST), the National Science Foundation (NSF), the National Oceanic and Atmospheric Administration (NOAA), the Federal Emergency Management Agency (FEMA), and, at the Director's discretion, the Federal Highway Administration (FHWA) to coordinate activities in furtherance of the program's objectives. Although the NWIRA authorization for appropriations expired in October 2007 and has yet to be reauthorized by Congress, these agencies continue to pursue measurable reductions in losses of life and property from windstorms per the original legislation.

As an initial step, the working group developed the *Windstorm Impact Reduction Implementation Plan*, which was sent to the Congress on April 5, 2006. The plan identified research activities needed to improve windstorm hazard mitigation and served as a guide for the Federal agencies to coordinate new and existing research to fill gaps in understanding, predicting, and forecasting windstorms. Building on that plan in a subsequent and related effort, and as part of a broader undertaking by the NSTC and SDR to identify and prioritize the Federal investments in science and technology needed to reduce future disaster losses, the NWIRP agencies contributed expertise to the development of a series of 15 implementation plans which, in addition to addressing other hazards, articulated necessary steps for reducing windstorm impacts. This series includes plans that prescribe specific science and technology actions that inform and support mitigating impacts of hurricane and tornado hazards, as well as the coastal inundation and flood hazards which hurricanes and other windstorms often produce. Published in 2008, these plans – the *Grand Challenges for Disaster Reduction*

Implementation Plans for *Hurricane*, *Tornado*, *Coastal Inundation*, and *Flood*[12] – are available online at the SDR website at http://www.sdr.gov/.

This document is the biennial report for Fiscal Years 2013 and 2014. The report provides examples of relevant Federal activities undertaken by the abovementioned agencies during the fiscal period (from October 2012 through September 2014). The examples herein reflect instances of collaboration and cooperation across levels of government, among Federal agencies, and with academia and the private sector. The sections of this report are organized in thematic sections according to the goals of: improved understanding of windstorms; impact assessment and reduction; and outreach. Within these sections, the report summarizes activities undertaken by:

- NIST, in support of research and development to improve codes, standards, and practices for design and construction of buildings, structures, and lifelines.
- NSF, in support of research in engineering, the atmospheric sciences, and social sciences to improve the understanding of the behavior of windstorms and their impact on buildings, structures, lifelines, and society.
- NOAA, in support of atmospheric sciences research to improve the understanding of the behavior of windstorms and their impact on buildings, structures, and lifelines.
- FEMA, in support of the development of risk assessment tools and effective mitigation techniques, windstorm-related data collection and analysis, public outreach, information dissemination, and implementation of mitigation measures consistent with the agency's all-hazards approach.
- FHWA, in support of research to improve the understanding of windstorm impacts on bridges and other highway structures to advance corresponding design, engineering, and construction standards.

4. Program Activities in Fiscal Years 2013 and 2014

4.1. Understanding Windstorms for Improved Predictions, Forecasts, and Warnings

Federal efforts to understand, predict, and forecast windstorms span a broad array of activities, from basic research funded by NSF and other agencies to severe weather warnings provided by NOAA. During Fiscal Years 2013 and 2014, NSF and NOAA supported research to advance the understanding of the physical processes that determine hurricane intensity, tornadogenesis, tornadic vortex structure, and other weather phenomena. NOAA activities included: research to improve observations of physical phenomena; development of novel data assimilation and forecasting techniques; and applications of observations, models, and forecasts.

4.1.1. Hurricanes

Through advances in satellite-based observations, supercomputers, and modeling, tremendous strides over recent decades have reduced average hurricane forecast track errors significantly. They are now about half of what they were 15 years ago.[13] Much of this progress is the result of advances in numerical weather prediction: that is, the use of computer models to extrapolate future weather based on current conditions and knowledge of atmospheric dynamics. Further advances in hurricane modeling will require the development of models that can accurately depict large-scale atmospheric flows, which are primarily responsible for steering hurricanes, while at the same time representing the finer-scale details of their inner cores, which influence their intensities.

Prediction of hurricane formation and intensification remains one of the most challenging aspects of atmospheric science, and efforts during the past decade have only yielded slight improvements in this area. Moreover, while statistical-dynamical models have provided the most accurate guidance for intensity prediction in recent years, experts at the National Hurricane Center (NHC) project that limitations inherent to these joint models will likely restrict their ability to derive further skill improvements from them to a range of 10 to 20 percent. Improvement beyond this ceiling must come from other applications. Within this context, high-resolution atmospheric modeling systems based on dynamical and ensemble approaches offer the best hope for significantly improving intensity forecasts.[14] Future progress in hurricane forecasting will depend upon the success of programs such as NOAA's Hurricane Forecast Improvement Program (HFIP) and others highlighted in the following section.

NOAA: Hurricane Forecast Improvement Program. The HFIP provides the basis for NOAA and other agencies to coordinate and align Federal research with that of the larger scientific community. The specific goals of the HFIP are to reduce the average errors of hurricane track and intensity forecasts by 20 percent within five years and 50 percent in approximately 10 years while extending the forecast period out to seven days with accuracy equal to today's five-day forecast. NOAA is making significant progress toward meeting these goals. In Fiscal years 2013 and 2014, HFIP continued to provide funding to numerous efforts within NOAA, other Federal agencies, and universities to support the development, testing, and evaluation of enhanced numerical prediction systems, better use of existing observation, evaluation of new potential observations, and support for improved forecasting techniques. Successful demonstration of the systems illustrated that the five-year track forecast goals should be achieved through the use of recently developed data assimilation systems, improved higher resolution models, and existing operational global models when run as an ensemble at 30 km resolution. Although there appears to have been significant progress in intensity prediction using high-resolution regional models, the past two seasons have provided relatively few challenging forecast scenarios. In addition, inner-core Doppler data are not consistently leading to forecast improvements, which likely reflect a continuing need to advance data assimilation systems and additional thermodynamic measurements in and around the hurricane core. However, hurricane model improvements have continued with the expectation that the five-year goal will be met in time for the calendar year 2015 hurricane season. Other notable successes under the program include the rapid deployment of models from research application to operational application and the development and testing of HFIP's Real-time Experimental Forecast System, which uses NOAA Research and Development High Performance Computing with real-time support to the NHC during the hurricane season.

NOAA: Joint Hurricane Testbed. During Fiscal Years 2013 and 2014, the U.S. Weather Research Program within the NOAA Office of Oceanic and Atmospheric Research (OAR) Office of Weather and Air Quality initiated the support of seven new projects conducted at the Joint Hurricane Testbed (JHT) in Miami, Florida – a NOAA testbed jointly managed by the NHC and the NOAA's Atlantic Oceanographic and Meteorological Laboratory (AOML). These projects involved joint testing of new techniques, applications, and model enhancements to improve the analysis and prediction of tropical cyclones. The JHT is designed to transfer more rapidly and smoothly new technology, research results, and observational advances by partnering researchers from the academic community, Federal agencies, and other groups with hurricane forecasters at the NHC throughout the entire project. These new projects represent a wide range of partners from academia (University of Miami, Florida International University, University of North Carolina, Florida State University, and University of Rhode Island) and U.S. government laboratories (NOAA's Geophysical Fluid Dynamics Laboratory, NOAA's Cooperative Institute for Research in the Atmosphere, and NOAA's Cooperative Institute for Meteorological Satellite Studies). During Fiscal Year 2014, earlier projects that had been recently completed were considered for operational implementation at either the NHC or the Environmental Modeling Center. Of these, six were accepted for implementation. Of particular note was the Tropical Cyclone Genesis Index, which will provide for the first time quantitative probabilistic guidance to assist NHC's Tropical Weather Outlook product.

NOAA/NASA: NOAA Intensity Forecasting Experiment. NOAA's Hurricane Research Division is in the midst of a multi-year experiment to improve hurricane-intensity forecasting. Developed in partnership with other parts of NOAA, the Intensity Forecasting Experiment (IFEX) is: collecting observations that span the tropical cyclone (TC) life cycle in a variety of environments for improvements in the initialization and evaluation of the next-generation Hurricane Weather Research and Forecasting operational model; developing and refining measurement strategies and technologies that provide improved real-time monitoring of TC intensity, structure, and environment; and improving the understanding of physical processes important to intensity change for a TC at all stages of its life cycle. Measurable progress is occurring in TC intensity forecasts as a result of these efforts, though much work remains. When possible, IFEX partners with other Federal agencies to accomplish a more complete sampling of TC structure and intensity. For example, during Fiscal Year 2014, IFEX partnered with the National Aeronautics and Space Administration (NASA) on the Hurricane and Severe Storm Sentinel (HS3) Project and the Office of Naval Research (ONR) Tropical Cyclone Intensification and Structure (TCI-14) Project in a multi-agency collaborative and coordinated study of changes in hurricane intensity and structure in the Atlantic basin. In combination, these agencies deployed multiple aircraft over a 45-day period during the peak of the Atlantic hurricane season to gather robust and complementary data sets. One of the noteworthy aspects of this collaboration was the use of NASA's remotely piloted Global Hawk scientific research unmanned aircraft system in a hurricane research capacity. It is particularly significant that the drone's 24-hour flight time gave scientists the ability to directly observe Hurricane Edouard as it changed over time in a way that conventional planes and satellites had not done before. In addition to demonstrating new skill in sampling the wind and thermodynamic structure of tropical cyclones, scientists have shed light on the processes tied to the rapid intensity and structure changes of Hurricane Edouard. These innovative interactions continued during Fiscal Years 2013 and 2014, with IFEX partnering with NASA and ONR in the NOAA Sensing Hazards with Operational Unmanned Technology (SHOUT) experiment to utilize a Global Hawk and the NASA WB-57 High Altitude Research Program for hurricane observation.

NSF: Hurricane Predictions and Predictability. NSF continues to invest in studies related to the basic science behind tropical cyclones and hurricanes. These studies range from the small-scale analysis of how sea spray and cloud particles affect the development and intensity of storms to the questions of the genesis of cyclones and how large-scale environmental factors affect their intensity and movement. Hurricane Sandy is a prime example of the tragic toll that hurricanes can take on lives and property. In Fiscal Year 2013, NSF-funded researchers at Pennsylvania State University began analysis of the evolution of Hurricane Sandy through its lifecycle and landfall. A particular focus of the research is on ensemble modeling, where the same model is run with slightly different conditions to determine how robust a solution is.

4.1.2. Tornadoes, Thunderstorms, and Other Severe Weather

Tornadoes and other severe weather such as thunderstorms also continue to claim lives and cause damage in the United States on an annual basis. Historic tornado events – such as the widest tornado ever recorded, which struck El Reno, Oklahoma, in May 2013 – cause havoc and terrorize communities. Investments over the last two decades have led to advances in the understanding of tornadogenesis and improvements in windstorm prediction and have enabled NWS to double the average lead time for tornado warnings to 13 minutes.[15] Yet even with the application of best-available science, the rate at which warnings result in false alarms remains high, at 70 percent[16]. To reduce this rate and further increase warning lead times for devastating windstorms such as tornadoes, the Federal agencies conducted an array of focused efforts during the fiscal period. Examples of agency efforts related to tornadoes, thunderstorms, and other severe weather follow below.

NOAA: Warn-on-Forecast. NOAA's National Severe Storms Laboratory (NSSL) is working with the NWS to develop a new vision for the warning decision process, which continues to evolve as scientists and engineers work toward integrating the next generation radar (e.g., rapid-scanning phased array radar and very rapid-scanning radar with data in less than one minute) and storm-scale numerical models to create a storm-scale prediction capability for the NWS. The NSSL continues to investigate various model parameterization schemes, along with techniques to improve model initialization through three- and four-dimensional data assimilation. In Fiscal Years 2013 and 2014, the NSSL continued to receive funding to support the Warn-on-Forecast (WoF) project. Within the next decade, the NSSL envisions operational units evaluating a WoF methodology, allowing forecasters to use thunderstorm-resolving computer models to guide severe weather warnings for tornadoes, winds, hail, and flash floods in the same way they do today with the current Doppler radar systems. In conjunction with the WoF project, NSSL is also demonstrating the capabilities of the Multifunction Phased Array Radar (MPAR), a program funded through NOAA's Office of Oceanic and Atmospheric Research with large technical contributions from industry and academia. The MPAR project was established to demonstrate the potential to simultaneously perform aircraft tracking, wind profiling, and weather surveillance with a single, phased array weather radar. NOAA's National Weather Radar Testbed centers on a phased array radar actively being tested and evaluated in Norman, Oklahoma. Evidence suggests that these enhancements to operational weather capabilities will lead to a more accurate warning system, increase lead time, and provide probabilistic information to the public to support risk-wise actions during severe weather events. The WoF program is being conducted in collaboration with the NOAA Earth System Research Laboratory Global Systems Division

(ESRL/GSD), the NWS Storm Prediction Center (SPC), and the NWS Forecast Office in Norman, Oklahoma (OUN).

NOAA/NSF: Verification of the Origins of Rotation in Tornadoes Experiment 2. In Fiscal Years 2009 and 2010, NSF and NOAA jointly supported the field phase of the Verification of the Origins of Rotation in Tornadoes Experiment 2 (VORTEX2), which was the largest and most ambitious tornado field study conducted to date. This $14 million effort involved nearly 100 scientists and students from 16 American universities and academic organizations, as well as forecasters from the NWS (including the SPC), Environment Canada, the Australian Bureau of Meteorology, and the Finnish Meteorological Institute. The VORTEX2 teams sought to better understand how, when, and why tornadoes form. Specifically, they investigated: the wind, temperature, and humidity conditions that result in the development of tornadoes within thunderstorms; the detailed structure of tornadic winds and their relationship to localized damage patterns; and the relationships between tornadoes, their parent thunderstorms, and the larger-scale environment. Continuing analysis of the VORTEX2 dataset in 2013 and 2014 is advancing the understanding of tornado formation, improving prediction, and offering prospects for increased warning lead-times. Ongoing analyses are pointing to a complex interplay of previously-recognized broad storm-scale rotation and far more subtle processes hinging on the detailed makeup and evaporative potential of precipitation particles within the rear-flank regions of supercell thunderstorms in triggering (or in some cases, apparently blocking) the spin-up of intense near-surface rotation on the tornado scale.

NOAA: Hazardous Weather Testbed. During Fiscal Years 2013 and 2014, NOAA's NSSL, SPC, and OUN continued joint testing of new techniques and applications for enhancing forecasts and warnings of hazardous weather – particularly thunderstorms and their attendant damaging winds, hail, and tornadoes. This work has been conducted in the NOAA Hazardous Weather Testbed (HWT), which is designed to accelerate the transition of promising new meteorological insights and technologies into advances in forecasting and warning for hazardous mesoscale weather events throughout the United States. The NOAA/NASA Geostationary Operational Environmental Satellite – R Series (GOES-R) Proving Ground also conducts experiments within the HWT and the University of Oklahoma's Center for Analysis and Prediction of Storms to provide key contributions each spring during peak severe weather season. Several collaborative experiments are conducted in the HWT each year, typically involving multiple external collaborators such as the National Center for Atmospheric Research, other universities, various NOAA agencies, private industry, and leading research scientists and forecasters from around the world.

NOAA: Toward Seasonal Prediction and Long-term Severe Weather Variability in a Changing Climate. In Fiscal Year 2012, the NWS SPC partnered with the NWS Climate Prediction Center, academia, and the NOAA NSSL to improve understanding of the links between large-scale climate variability and windstorm and tornado activity. The SPC maintains the longterm observational record of severe weather in the United States and a result of initial interactions, NOAA published a "State-of-the-Science Fact Sheet" in Fiscal Year 2013 titled *Tornado Activity and Climate Variability and Change*. While climate change can be expected to increase temperature and humidity in the lower atmosphere, it may also decrease wind shear. Since both factors are required to form tornadoes, NOAA research is focused on the prediction of these local factors as well as quantification of the role of more-remote climate signals.

NSF/NOAA: Predictability of Severe Convective Storms. In Fiscal Year 2013, NSF and NOAA researchers conducted the Mesoscale Predictability Experiment (MPEX). The goal of the experiment was to determine whether additional observations of weather conditions upstream of the U.S. Great Plains could improve predictions of severe convective storms, including tornadoes and straight-line wind producing events. The NSF/NCAR GV research aircraft deployed hundreds of instruments known as dropsondes in order to retrieve profiles of wind, moisture, and temperature of the atmosphere below the flight level of the aircraft. Weather balloons were also launched from the ground to supplement the aircraft observations. Analysis of the data is ongoing, and should allow researchers to determine whether it is observations or numerical models that are the limiting factor for prediction of these severe storms.

NSF: Tornado Theory and Observations. NSF continues to fund research in the fiscal period making use of the extraordinary efforts of the VORTEX2 field campaign that was originally conducted in Fiscal Years 2009 and 2010. In addition, new lines of research regarding the basic theory of tornadoes and ways to better observe them are now being conducted. In one case, a group of researchers at the University of Oklahoma are studying how tornado debris appears to weather radars. The recent upgrade to the Nation's WSR-88D radar system to retrieve polarimetric variables has allowed scientists to better determine the sizes and shapes of objects that the radars are detecting. Researchers are now investigating how items commonly lofted by tornadic winds, such as branches, pieces of wood, and small plastic and metal objects, are observed by weather radar. By better understanding what the radar is observing, weather forecasters and emergency managers can more promptly respond to recent tornado damage and provide better warnings to those along a tornado's path.

4.2. Assessing and Reducing Windstorm Impacts

Federal efforts to assess and reduce the impacts of windstorms cover a wide range of activities, including: studying the performance of buildings, structures, and infrastructure during and after windstorms; research, development, and technology transfer of windstorm modeling, simulation, risk assessment, and loss estimation techniques; development of cost-effective windstorm-resistant materials and systems; and information dissemination and outreach to a wide range of stakeholders in the engineering and construction industries, government, and the general public (see Section 4.3 for additional information on outreach activities).

During Fiscal Years 2013 and 2014, NSF and NOAA supported research in areas of simulation of hurricane and tornado wind fields and the understanding of tornado, hurricane, and wind-driven rain effects on buildings. NIST activities included development of improved tools for estimating wind hazards and estimation of wind loads on structures. NIST and FEMA conducted post-windstorm studies, and the results of such prior and current studies led to changes in building codes to improve windstorm resistance. NOAA and FEMA supported the continued development of hurricane loss estimation models. FHWA activities included real-time monitoring of weather conditions and bridge response to winds and physical and computational modeling of wind effects on highway structures. NSF also supported research related to resilience to natural hazards.

4.2.1. Buildings and Structures

NIST: Wind Engineering. NIST's wind engineering research is focused on developing the measurement science, tools, and methodologies necessary to substantially improve: 1) wind hazard estimation through better extreme wind databases and maps; 2) the estimation of wind loads on structures, through the development of database-assisted design techniques incorporating wind directionality effects and prediction of structural responses to these loads; and 3) science-based methodologies for aerodynamic simulation and measurements, for determination of wind loads on buildings and structures. Progress during Fiscal Years 2013 and 2014 included: continued development of new design wind speed maps intended for use in the American Society of Civil Engineers (ASCE) Standard on Minimum Design Loads for Buildings and Other Structures (ASCE 7); conversion of an existing smooth flow wind tunnel test section to a boundary layer test section; and initiation of computational fluid dynamics research to study computationally efficient means for simulation of atmospheric turbulence needed for accurate simulations of wind loads on buildings,[17,18] to be supported by experiments conducted in the new boundary layer wind tunnel test section.

FEMA: Improving Wind-Resistant Provisions. During the fiscal period, FEMA continued to work with its partners to develop and incorporate high-wind-resistant provisions and requirements in the Nation's model building codes and standards. Working with other Federal agencies, state and local governments, building regulators, building industry groups, and other entities, FEMA advocated for specific changes to increase wind-resistant requirements of the following building and structural codes: International Building Code (IBC), International Residential Code (IRC), International Code Council Storm Shelter Standard (ICC 500) and Residential High Wind Standard (ICC 600), and other industry regulations such as ASCE 7. As a result of these efforts, buildings are being built stronger, reducing the risks of death, injury, and property loss from high-wind storms.

As a result of the 2011 spring tornadoes Mitigation Assessment Team (MAT) research, FEMA proposed two code changes, which were approved for publication in the 2015 IBC. The code changes call for mandatory storm shelters in high-wind regions for schools and first-responder facilities as follows:

- In areas where the shelter design wind speed for tornadoes is 250 mph, all new kindergarten through 12th grade schools with 50 or more occupants in total per school shall have an ICC 500-compliant storm shelter.
- In areas where the shelter design wind speed for tornadoes is 250 mph, all new 911 call stations, emergency operation centers, and fire, rescue, ambulance and police stations shall have an ICC 500-compliant storm shelter.

FEMA: Mitigation Assessment Team Evaluations Following Major Hurricanes, Tornadoes and Windstorms. In Fiscal Year 2013, FEMA MATs responded to both Hurricane Isaac in Louisiana and Hurricane Sandy in New Jersey and New York. FEMA deployed a MAT to investigate the damage and provide technical assistance to the affected communities through the Joint Field Offices established in response to these events. Examples of some of the products conceived as a result of recent studies include two Recovery Advisories (RA) for Hurricane Isaac; and seven RAs and 3 fact sheets from the Hurricane Sandy MAT:

- RA 1 – Improving Connections in Elevated Coastal Residential Buildings
- RA 2 – Reducing Flood Effects in Critical Facilities
- RA 3 – Restoring Mechanical, Electrical, and Plumbing Systems In Non-Substantially Damaged Residential Buildings
- RA 4 – Reducing Interruptions to Mid- and High-Rise Buildings During Floods
- RA 5 – Designing for Flood Levels Above the Base Flood Elevation (BFE) After Hurricane Sandy
- RA 6 – Protecting Building Fuel Systems from Flood Damage
- RA 7 – Reducing Flood Risk and Flood Insurance Premiums for Existing Residential Buildings in Zone A
- Fact Sheet 1 – Cleaning Flooded Buildings
- Fact Sheet 2 – Foundation Requirements and recommendations for Elevated Homes
- Fact Sheet 3 – Building Science Support and Code Changes Aiding Sandy Recovery

The Hurricane Isaac MAT was charged with evaluating hurricane damage with focus on performance of post-Hurricane Katrina construction and reconstruction efforts. This included critical facilities affected by the storm, evaluating the performance of electrical distribution and communication facilities, and investigating claims of wind damage in newer buildings. The MAT's observations, conclusions, and recommendations are presented in FEMA P-938, *Mitigation Assessment Team Report – Hurricane Isaac in Louisiana: Building Performance Observations, Recommendations, and Technical Guidance.*[19]

The Hurricane Sandy MAT investigated the performance of building codes and standards; flood protective measures; residential construction; critical facilities and key assets; and mechanical, electrical, and plumbing systems. The final MAT report (FEMA P-942: *Mitigation Assessment Team Report – Hurricane Sandy in New Jersey and New York; Building Performance Observations, Recommendations, and Technical Guidance*[20]) was released in Fiscal Year 2014 and presents the conclusions and recommendations derived from the field observations with regards to key engineering concepts, codes and standards, mitigation measures and considerations that can be used in the planning and recovery process to help minimize future damage to structures and their related utility systems.

In Fiscal Year 2014, FEMA also released a Formal Observation Report, FEMA P-1020[21], presenting the observations, conclusions, and recommendations in response to field investigations conducted after the EF-5 tornado that impacted the towns of Newcastle and Moore, Oklahoma, on May 20, 2013. The post-storm investigation team focused its efforts on safe rooms and storm shelters in the path of the tornado in order to analyze their performance, functionality, and use.

NIST/NOAA: Preliminary Reconnaissance of the Newcastle-Moore Tornado. At the time of the Newcastle-Moore tornado, NIST was nearing completion of its technical investigation of the 2011 Joplin tornado. The objectives of the Newcastle-Moore Tornado Preliminary Reconnaissance, conducted in collaboration with NOAA, were therefore limited to identifying information that might be relevant to aspects of the Joplin tornado investigation, including collection of data and information on: 1) the performance of the emergency communications systems immediately prior to and during the tornado; and 2) the response of critical and educational facilities to the tornado, including emergency operations, the physical performance of the buildings and designated safe areas, and life safety outcomes. One of the key observations described in the preliminary reconnaissance report[22] was that designated safe areas at one of the impacted elementary schools did not provide life safety protection. Seven schoolchildren died and several others were injured when part of the building's designated safe area, in which they were taking shelter, collapsed.

NIST: Technical Investigation of the Joplin Tornado. Following the May 22, 2011, tornado that devastated the city of Joplin, Missouri, NIST sent four engineers to the area from May 24-28 to conduct preliminary reconnaissance. This was the deadliest single tornado since official record keeping began in 1950, causing 161 fatalities. Based on analysis of the data collected and other criteria required by law and regulation, the Director of NIST established a team (including staff from NIST and NOAA) under the National Construction Safety Team (NCST) Act on June 29, 2011, to proceed with a more comprehensive study of the disaster. The final report of the technical investigation.[23] completed in Fiscal Year 2014, includes 47 findings that support 16 recommendations in the following broad areas: improving understanding of tornado hazard characteristics and associated wind fields; improving performance of buildings, shelters, designated safe areas, and lifelines; and improving performance of emergency communications systems and public response. Specifically, the NIST report includes calls for: 1) developing national performance-based standards for tornado-resistant design of buildings and infrastructure, as well as design methods to achieve those standards; 2) developing improved standards, codes and guidelines for siting, design, installation, and operation of tornado shelters, and installation of many more shelters in tornado-prone areas; and 3) creating national codes and standards for clear, consistent and accurate emergency communications and then ensure that emergency managers, the NWS, and the news media in local communities have a joint plan for delivering those messages quickly and persuasively during tornadoes.

NIST: Implementation of Recommendations from the Joplin Tornado Investigation. Following completion of its technical investigation of the Joplin Tornado, NIST initiated a new project to begin implementing the 16 recommendations identified in the final report.[24] Progress during Fiscal Year 2014 included approval of a proposal submitted to the ASCE to develop a new standard on methods of wind speed estimation in tornadoes. The scope of the new standard will include updating and improving the EF scale for rating tornadoes, development of standardized methods for radar data analysis, forensic engineering, and other wind speed estimation techniques, and minimum data archival requirements to support wind speed estimates. The new ASCE committee charged with developing this standard is co-chaired by a meteorologist from NOAA/NWS and a wind/structural engineer from NIST. A multi-year effort to develop new tornado hazard maps, for use in performance-based design of buildings and structures to resist tornadoes, was also initiated in Fiscal Year 2014.

FEMA/FHWA/NIST/NSF: U.S.-Japan Panel on Wind and Seismic Effects. FEMA, FHWA, NIST, and NSF all participate on the U.S.-Japan Cooperative Program in Natural Resources Panel on Wind and Seismic Effects. NIST also chairs and provides a technical secretariat for the panel. The panel serves as a mechanism for the exchange of technical data, information, and researchers, as well as the coordination of joint research on wind and seismic-related topics of mutual interest to the U.S. and Japan. Task Committees addressing wind and storm surge hazards include Task Committee "D" (Wind Engineering), Task Committee "G" (Transportation), and Task Committee "H" (Storm Surge and Tsunami). The Task Committee "D" group has held five international workshops and the Task Committee "G" group has held 30 such workshops. During Fiscal Year 2013, the U.S. side hosted a panel meeting at NIST in

Gaithersburg, Maryland. The U.S. side also hosted a bridge engineering workshop in Portland, Oregon, and attended a bridge engineering workshop in Tsukuba Science City, Japan, during Fiscal Year 2013.[25]

NIST/NSF: R&D Roadmap for Windstorm and Coastal Inundation Impact Reduction. NIST completed and published an R&D roadmap for windstorm and coastal inundation impact reduction[26] during the current reporting period. The roadmap includes a vision for windstorm and coastal inundation resilient communities, identifies grand challenges, and provides descriptions of 30 priority R&D topics in areas of hazard characterization, loads and effects, resistance to loads, and performance-based design for wind, storm surge, and tsunami hazards. Input for the roadmap was acquired through a pair of workshops (held in Reston, Virginia, in Fiscal Year 2012, and Miami, Florida, in Fiscal Year 2013), the first of which was jointly sponsored by NSF. Over 100 industry leading practitioners, academics, and representatives from key government agencies participated in the two workshops. Federal government participants included representatives from the U.S. Department of Housing and Urban Development and the U.S. Geological Survey, in addition to FEMA, FHWA, NIST, NOAA, NSF, and OSTP.

NSF: RAPID Awards: Collection of Perishable Data. NSF has a standing program to provide funding to university faculty members to collect perishable data from the field after a significant windstorm event. During Fiscal Years 2013 and 2014, the institutions that were granted funds for collecting data after Hurricane Sandy completed their work. The data include level and impact of storm surge on residential structures and bridges in the New Jersey and New York coastal areas, power outages in New York City and surrounding areas, and impact of wind in inland areas. The institutions that participated in the research included: New York University, CUNY City College, Princeton University, Drexel University, Stevens Institute of Technology, Louisiana State University, University of Washington, and the University of Notre Dame. The collected data are housed on the web sites of the institutions and are available to principal investigators and the public. A similar effort of collecting data after Typhoon Haiyan that devastated islands of the Philippines was pursued by the University of North Carolina-Charlotte. In addition, faculty and students from University of Oklahoma, Mississippi State University, and University of Alabama documented perishable data and modes of collapse of buildings following the tornadoes that impacted the Newcastle, Moore, and El Reno, Oklahoma, areas in May 2013. This collection of field data after the tornado has led to significant changes in the building codes of the city of Moore, Oklahoma. Faculty and students from Lyndon State College were able to perform detailed ground and aerial surveys of the damage track caused by the Newcastle-Moore and El Reno tornadoes.

NSF: Planning Grant for Industry/University Cooperative Research Center (I/UCRC) for Windstorm Hazard Mitigation. NSF has provided a collaborative grant to Texas Tech University, Louisiana State University, and the University of Alabama to formulate a partnership center for research and development on windstorm hazard mitigation between the academic institutions and industry. The grant is for planning purposes and allows the group to meet and delineate the type of research that the industry would like the researchers to pursue. The goal is to produce synergy between academic researchers and industry personnel to use basic research and facilities to develop meaningful results that the industry can use. The plan of the center is to continue beyond funding from the NSF.

NSF: Awards for Response of Tall Buildings in High Winds. As buildings are planned and designed to be greater and greater in height, NSF is funding awards to assess the response of these tall structures in high winds and to create improved aerodynamic shapes to reduce drifts. Very tall buildings in the range of 100 stories or more can sway in high winds, and this swaying can be felt by people – making them uncomfortable. The optimum aerodynamic shape of a building can reduce swaying and the drift of the building; in addition, aerodynamic shapes can reduce wind-induced loadings and make very tall buildings more resilient to winds and potentially more economical. Awards were given in the fiscal period to Northeastern University and the University of Notre Dame to study emerging problems related to tall buildings.

NSF: Hurricane Wind and Surge Effects on Buildings and Bridges. NSF has provided grants to the University of Notre Dame and the City University of New York – City College to pursue fundamental research in understanding the impact of hurricanes on shoreline buildings and bridges.

These impacts are due to high winds as well as storm surge. In the research projects, loads are modeled in computer simulation to understand their impact on the coastal buildings and bridges. Responses of buildings to storm surge and waves are modeled using a structural mechanics approach, and responses will be compared with the damage data collected after Hurricane Sandy. The computer simulation model will be refined to duplicate the field data to the extent possible, and in addition, the computer simulation of storm surge will predict scour that the surge can cause around bridge piers. Another grant provided by NSF is developing rapid deployment of self-contained instrumentation that can measure loads on bridges during a hurricane. The Florida Department of Transportation is working closely with University of Florida faculty and students to facilitate rapid deployment of the instruments. This early-career grant will establish storm surge and wave impact on bridge structures in hurricanes using wireless smart sensors assisted by advanced cyberinfrastructure. The results of these studies have potential of improving coastal buildings and bridges to resist hurricane forces in the future.

4.2.2. Bridges and Highways

FHWA: Monitoring of Wind Conditions and Structural Performance. During the fiscal period, the FHWA Office of Infrastructure Research and Development monitored winds and structural performance at selected bridge sites to establish and characterize site-specific wind conditions and the responses of the bridges to varying wind conditions. The sites of two major long-span, cable-supported bridges were monitored during the reporting period – the Deer Isle- Sedgwick Bridge (Maine) and the Hale Boggs Bridge (Louisiana). Data were collected on a continuous basis; however, few major storm events were observed. The data gathered during this monitoring period are being used for: the calibration of numerical tools and predictive methods; the improvement of physical modeling techniques in the laboratory; the evaluation and validation of new designs; and the development and assessment of mitigation measures.

FHWA: Full Scale Measurement of Structural Dynamic Properties. In an ongoing program to characterize the dynamic properties of bridge stay cables – which are critical to structural performance and aerodynamic stability – the analysis of data from full-scale, forced vibration tests on selected major structures continued during Fiscal Years 2013 and 2014. Tests had been previously performed on the Zakim (Massachusetts), Emerson (Missouri), and Penobscot Narrows (Maine) bridges. During this reporting period, a technical report[27] on the Penobscot study and a technical paper[28] on the Emerson study were published. The report documents the behavior of large, ungrouted bridge cables and illustrates the effectiveness of viscous dampers as a mitigation measure. The paper summarizes the behavior of grouted and ungrouted cables as well as the effectiveness of cross ties as a mitigation measure. Information collected during the tests has enabled further evaluation of design details and assessment of the effectiveness of various mitigation measures such as dampers, cross ties, and aerodynamic surface treatments. The information also serves as a benchmark that will be useful during later inspections to assess the bridge's structural condition and overall health.

FHWA: Physical Modeling of Wind Effects on Highway Structures. In collaboration with the National Research Council of Canada (NRCC) and in cooperation with the University of Stavenger in Norway, the University of Bristol in the United Kingdom, and the design consultancy firm RWDI in Canada, wind tunnel tests were performed (during the previous reporting period) at NRCC on a full-scale section model of a bridge stay cable. Tests were performed in simulated wind conditions over a broad range of wind speeds to evaluate the influence of wind speed, turbulence intensity, wind and cable orientation, cable roundness, and aerodynamic surface treatment on overall performance and aerodynamic stability. During Fiscal Years 2013 and 2014, a comprehensive technical report[29] and four technical papers[30,31,32,33] were published. As noted in the report, the imperfect shape of High Density Polyethylene Pipe (HDPE) used on bridge stay cables can play an important role in the aerodynamic stability and performance of cables. Additional tests have been planned in the next reporting period to further investigate this area of wind and cable interaction. The design of these tests requires information regarding the roundness of typical HDPE used in bridge construction. To obtain this information, detailed field measurements must be undertaken at select bridge sites and an automated robotic device has been designed and fabricated for this purpose. Large Variable Message Signs can and have exhibited sensitivity to the effects of wind and truck gust loadings resulting in performance problems and structural failures. To investigate this, wind tunnel testing in the FHWA Aerodynamics Laboratory and computational modeling in the Argonne National Laboratory are ongoing. During this reporting period, wind force measurements on a 1:10 scale model of a representative message sign panel were completed for a wide range of vertical wind angles. Traffic signal standards and high light towers have also exhibited sensitivity to the effects of wind and structural failures have been observed. Research on this subject has been conducted under the National Cooperative Highway Research Program (NCHRP), and American Association of State Highway and Transportation Officials (AASHTO) standard specifications for these structures have been significantly updated. To evaluate and further build upon these improvements in design, wind tunnel testing in the FHWA Aerodynamics Laboratory is ongoing. During this reporting period, wind force measurements on multi-sided, tapered cylinders have resumed and tests were completed on 18 sectional models with representative geometries.

FHWA: Numerical Modeling of Wind Effects on Highway Structures. During the reporting period, an existing interagency agreement has continued between the U.S. Department of Transportation and the U.S. Department of Energy for modeling the effects of natural hazards, such as windstorms, hydraulics, and flooding, on infrastructure. In addition, a second agreement was initiated to continue this collaboration and broaden the scope of research. These agreements have enabled FHWA to utilize high-performance computing and support staff from the Argonne National Laboratory (ANL). Resources located at the Transportation Research and Analysis Computing Center (TRACC) are being used to model wind effects on highway users as well as highway structures. To evaluate performance of our facility and to enhance simulation capability, a detailed Computational Fluid Dynamics (CFD) model of the FHWA Aerodynamics Laboratory at the Turner Fairbank Highway Research Center (TFHRC) – including the large wind tunnel – was completed and a technical report[34] has been published. Research has continued in this area to further refine and streamline the model with the objective of developing an effective computational framework, or "virtual wind tunnel," that will complement physical testing facilities. Special techniques have been developed by ANL staff to couple CFD software with Computational Structural Mechanics (CSM) software to enable study of wind and structure interaction problems. These emerging tools have been used to develop and refine models to study the interaction of wind with inclined bridge stay cables as well as wind (or truck-induced gust) interaction with large message sign and signal structures. FHWA and ANL staff have also refined skills with using "moving mesh" capabilities of CFD software so that objects, such as trucks, can be moved through still air or wind fields to study fluid and structure interaction. This complex meshing has been used to develop a computational model for studying the wind "shielding" effects of large bridge towers on trucks crossing elevated bridge structures. These CFD tools, as well as multi-physics modeling capabilities, were used to study the impact of salt spray from trucks, moving either through still air or a wind field, on weathering steel girders of highway overpasses. The first phase of this work established a modelling philosophy, which was found to include many assumptions regarding the initial size and velocity of spray droplets coming from tires. These results have been documented in an interim report,[35] and to validate the aforementioned assumptions, field measurements were conducted and completed. During the previous fiscal period of 2011 and 2012, finite element modeling was performed using the commercial program SAP2000 to study the efficacy of various approaches for mitigating wind-induced vibration of bridge stay cables. In the current reporting period, this study has been completed and a comprehensive laboratory report[36] has been published.

FHWA: Development of Design Guidelines and Specifications. Research continued in Fiscal Years 2013 and 2014 on the issue of wind- and rain-induced vibration of bridge stay cables. The FHWA draft guidelines document on the aerodynamic design of bridge stay cables was further updated, which will provide guidance on the aerodynamic design of cables and cable networks on major new highway bridges and for the retrofit of existing structures to mitigate wind-induced vibration problems. Results of research on bridge cables have been made available to the Post-Tensioning Institute's DC-45 Cable-Stayed Bridge Committee for consideration during periodic updates of their guide specification on cables. During this reporting period, FHWA also worked closely with the Transportation Research Board on the project NCHRP 20-07/Task 325 for updating AASHTO LRFD wind load provisions. The final study report and recommendations were submitted to the AASHTO SCOBS, Technical Committee on Load Distribution (T-5), for consideration in making changes to the code provisions on wind load.

4.2.3. Power Grid, Aviation, and U.S. Space Centers

NOAA: High-Resolution Rapid Refresh Model (HRRR) and the Rapid Refresh Model (RAP). There is a longstanding need for a higher-resolution weather models to increase accuracy of local forecasts for emergency managers, air traffic managers, renewable energy producers, wildfire managers, and others. To fulfil this need, scientists at the NOAA ESRL/GSD developed a radar-initialized, storm-resolving, high-resolution, hourly-updating assimilation and modeling system known as the High Resolution Rapid Refresh (HRRR) model to advance the prediction of weather changes in the environment. The HRRR provides hourly weather updates at 3 km resolution (i.e., a 3.5 square-mile area), which increases the protection of lives and property, helps to improve the efficiency and effectiveness of wind and solar energy of the U.S. energy sector, and provides air traffic managers with rapidly updated projections of developing severe weather. During Fiscal Year 2014, NOAA ESRL/GSD and NOAA National Centers for Environmental Prediction scientists worked together to transition the HRRR model into NWS operations, providing forecast products to users beginning on September 30, 2014. Version 2 of the HRRR is scheduled for release in late Fiscal Year 2015. The Rapid Refresh Model (RAP) is a coarser version of the HRRR assimilation and modeling system, providing forecasts for the North American domain and updated hourly. The RAP provides the boundary conditions used to execute the HRRR model. The initial version of the RAP was delivered to NWS operations in Fiscal Year 2014. A new version of the RAP, planned for delivery to NWS operations at the end of Fiscal Year 2015, will include a larger domain and forecast out to nearly 30 hours.

NOAA: Hazardous Weather System for U.S. Space Centers. Operations at U.S space centers are heavily dependent on severe weather conditions and windstorm forecasts. Safety regulations restrict work on tall gantries, shuttle transport, refueling, and other operations during high wind events. Winds may also impact safety following rocket blasts, fuel spills, and other accidents by carrying toxins from an accident zone farther afield. The NOAA ESRL/GSD developed and implemented a state-of-the-art analysis and forecasting system as part of the Range Standardization and Automation program. A new version of the model was delivered to the Range in Fiscal Year 2014. The latest upgrade of the system – a triple-nested, three-dimensional analysis with a medium-range forecast – is currently undergoing on-site testing at Vandenberg Air Force Base in California.

NOAA/NSF: Reliability-Based Hurricane Risk Assessment for Offshore Wind Farms. Meeting the national goal of generating 20 percent of total energy needs from renewable source by 2030 will require placing wind farms offshore. These wind farms located on the Gulf and East Coasts are vulnerable to hurricanes, and in order to make offshore wind farms viable economically, it is necessary to have knowledge of how hurricane hazards can impact the fragility of wind turbines. NOAA AOML and Florida State University scientists are working with the Department of Energy (DOE) on evaluating the impact of hurricane winds on the turbine design and performance. DOE is providing support to collect and analyze vertical wind profiles from dropsondes within tropical cyclones affecting wind farm lease locations.

4.2.4. Community Services and Resilience

NOAA: State of Florida Public Hurricane Loss Model. NOAA continued to work with the state of Florida during the fiscal period on the Florida Public Hurricane Loss Model (FPHLM). The FPHLM is an open, transparent computer model that is used by the Florida Office of Insurance Regulation to provide a baseline for evaluating rate change requests for windstorm insurance. The FPHLM is the first such model that enables all of the results and details from the modeling approach to be open to scrutiny. The model's engineering component estimates damage to residential structures within Florida zip codes and an actuarial component then estimates the insured loss. The average annual loss is then estimated statewide for every zip code in Florida. The FPHLM was originally approved by the Florida Commission on Hurricane Loss Projection Methodology in August 2010 and successfully passed the state certification process initiated in late 2012. NOAA researchers at AOML and the Cooperative Institute for Marine and Atmospheric Studies partnered with the Florida International University researchers in the fiscal period to update and maintain the wind model within the FPHLM.

FEMA: Maintain Hazus-MH Hurricane Loss Estimation Model. FEMA developed and maintained the Hazards – U.S., Multi-Hazard model (Hazus-MH), a nationally applicable standardized methodology that contains models for estimating potential losses from multiple hazards, including hurricane winds. Hazus-MH uses geographic information system (GIS) technology to estimate physical, economic, and social impacts of disasters and graphically illustrates the limits of identified high-risk locations due to a region's hazards. Users can then visualize the spatial relationships between populations and other more permanently fixed geographic assets or resources for the specific hazard being modeled, a crucial function in the pre-disaster planning process. During the reporting period for Fiscal Years 2013 and 2014, the Hazus program was undergoing transition activities within the FEMA Risk MAP Program with little major activity concerning Hazus functionality. Hurricane-related activity during this time period included the release of Hazus 2.1 Patch 2 in February 2014, which updated Hazus compatibility with the HURREVAC model. The patch allows Hazus users to resume real-time downloads of HURREVAC products for integration with the Hazus hurricane model and better estimate potential structural damage and economic losses for a hurricane approaching the U.S.

NSF: Improving Economic Resilience on the U.S. Gulf Coast Communities to Coastal Hazards. This project funded through NSF in the fiscal period studies why communities with similar socio-economic compositions and facing identical disasters have different vulnerabilities and recovery rates. The goal is to explore the relationship of economic vulnerability, resilience, and recovery in three coastal counties in Mississippi. The research will identify variables that contribute to the recovery of these communities after disasters. It will directly benefit Gulf Coast communities in their efforts to increase resilience to tropical cyclones. Among the many benefits of the project are: increasing local participation in decision making regarding tropical cycles; aiding local policymakers in the design and deployment of effective mitigation and economic development policies; and constructing a web-based Spatial Decision Support System.

NSF: Quantifying Disaster Resilience of Critical Infrastructure-based Societal Systems with Emergent Behavior and Dynamic Interdependencies. This NSF-funded project during the reporting period is creating a way to measure the resilience of critical infrastructure-based societal systems (CISSs) that are necessary for community functioning. A CISS is comprised of interdependent buildings that together serve a community function and are dependent on networks of critical lifelines. Examples include such infrastructure dependent systems as school districts, university campuses, central business districts, and health care systems, which is the focus of this research. The analysis requires dynamic fault-tree and modeling of complex, adaptive systems with time-varying states and changes. Results of this research can be used to improve, build, and maintain communities that are more likely to withstand disruption or disaster. The research will also inform practicing health care managers and emergency planners.

4.3. Outreach

Reducing susceptibility to and destruction from severe windstorms requires action beyond the purview of the Federal government. Zoning laws, building codes, and jurisdictional resources central to promoting pre-disaster mitigation and fostering disaster-resilient communities are controlled by state and local authorities. Ultimately, state, county, and municipal officials must implement the sound land use practices, informed urban planning techniques, prudent development and reconstruction decisions, and effective disaster response and recovery strategies that underpin resilient communities. Windstorm disaster risk reduction is equally critical at the individual and family levels and must be facilitated throughout community social networks.

History shows that a lack of situational awareness and preparation amongst citizens are common threads among many major wind-related disasters. By knowing one's vulnerabilities and what actions should be taken to lessen them, the public can reduce the effects of windstorms. Disaster preparedness significantly improves one's ability to respond to the consequences of a windstorm hazard event – this means having emergency plans in place concerning what to do and where to go if a windstorm warning is issued or a hazard is observed. This type of community resilience, especially in vulnerable social populations, can be enhanced through a variety of means, including outreach and awareness programs and partnerships among Federal and state agencies and the other members of the local community. Numerous informational and educational materials are distributed each year to promote the protection of individuals and property from high-wind events, including hurricanes, tornadoes, and straight-line winds from thunderstorms. Selected examples of such activities are highlighted below.

NOAA: Weather-Ready Nation. In the fiscal period of this report, the Weather-Ready Nation (WRN) initiative continued to make progress. The purpose of WRN is, first and foremost, to save more lives and livelihoods. By increasing the Nation's weather-readiness, the country will be prepared to protect, mitigate, respond to, and recover from weather-related disasters such as severe windstorms. Society's ability to prepare for natural disasters requires a societal response equal to the risk, which is why the NWS is leveraging its vast nationwide network of partners and incorporating new ones who are beginning to share the vision of building a weather-ready nation. Partners in this program include: other Federal agencies, emergency managers, researchers, the media, the insurance industry, non-profits, and the private sector. Building a weather-ready nation starts with these internal actions but requires the action of a vast nationwide network of partners, including: other U.S. government agencies and emergency managers, researchers, the media, insurance industry, non-profits, the private sector, the weather enterprise, and more. Through a series of symposiums, the national dialog engages these partners in assessing why the Nation is experiencing such extreme impacts. Other service enhancements provided by WRN include: cataloging local impacts from weather to give the public and local governments more information about what actions they need to take; providing better explanations about differences between the forecast and the range of possibilities to help decision-makers plan for any possible scenario; mapping flood threats to allow emergency managers keep the public out of areas that are expected to flood; improving the distribution of weather information by keeping up with new communication technologies and social media; and working to achieve local weather-readiness where everyone gets a clear, understandable message when weather hazards such as severe windstorms are imminent and knows what to do to protect themselves.

FEMA: Education, Outreach, and Information Dissemination. The FEMA website, http://www.fema.gov serves as the Nation's portal to emergency and disaster information. FEMA publications related to wind and coastal surge hazards are available for free on the web on FEMA's media library, the Google Books website, and MADCAD. Dozens of presentations were given at conferences and other forums highlighting the importance of both hurricane mitigation and safe rooms. FEMA has partnered with communities and community organizations to teach design professionals and local officials the concepts of hurricane mitigation and safe room design. FEMA has maintained its partnership with external organizations such as the Disney Corporation and the Federal Alliance for Safe Homes (FLASH), which have been operating an interactive exhibit titled "Storm Struck: The Tale of Two Homes" at Epcot at the Walt Disney World Resort Innoventions Pavilion for the last several years with over 700,000 annual visitors. The FLASH attraction allows visitors to experience a severe weather incident, such as a hurricane, with the goal of teaching the guests about cutting-edge technology used to protect homes. FEMA sponsored and provided technical direction for the National Building Museum's "Designing for Disaster" Exhibit, which opened May 11, 2014. Visitors to the exhibit explore new approaches in design and engineering to protect life and property against a range of natural hazards, including hurricanes and tornadoes. The exhibit includes a partially deconstructed safe room following the design plans in FEMA P320. FEMA also sponsored curriculum kits that teach students in grades 7-9 about planning and designing in preparation for natural disasters.

FEMA: Development of Technical Guidance Materials on State-of-the-Art Wind-Resistant Design and Construction Methods. Each year, many thousands of publications dealing with wind hazards are ordered and distributed by FEMA. For example, FEMA's safe room guidance publications are among the most widely downloaded and distributed documents by FEMA's library and publications warehouse (FEMA P-320: *Taking Shelter From the Storm: Building a Safe Room For Your Home or Small Business* and FEMA P-361: *Design and Construction Guidance for Community Safe Rooms*). During the fiscal period, almost 12,000 copies of FEMA P-320 and over 3,400 copies of FEMA P-361 were distributed to organizations across the U.S. Other recent examples include:

- As part of the MAT Program, published the following reports and recovery advisories:
  - FEMA P-938, Mitigation Assessment Team Report – Hurricane Isaac in Louisiana: Building Performance Observations, Recommendations, and Technical Guidance (March 2013)
  - Hurricane Isaac RA 1 – Minimizing Wind and Water Intrusion by Covering the Underside of Elevated Buildings
  - Hurricane Isaac RA 2 – Minimizing Flood Damage to Electrical Service Components
  - FEMA P-942: Mitigation Assessment Team Report – Hurricane Sandy in New Jersey and New York; Building Performance Observations, Recommendations, and Technical Guidance (November 2014)
  - Hurricane Sandy RA 1 – Improving Connections in Elevated Coastal Residential Buildings

- o Hurricane Sandy RA 2 – Reducing Flood Effects in Critical Facilities
- o Hurricane Sandy RA 3 – Restoring Mechanical, Electrical, and Plumbing Systems In Non-Substantially Damaged Residential Buildings
- o Hurricane Sandy RA 4 – Reducing Interruptions to Mid- and High-Rise Buildings During Floods
- o Hurricane Sandy RA 5 – Designing for Flood Levels Above the BFE After Hurricane Sandy
- o Hurricane Sandy RA 6 – Protecting Building Fuel Systems from Flood Damage
- o Hurricane Sandy RA 7 – Reducing Flood Risk and Flood Insurance Premiums for Existing Residential Buildings in Zone A
- o Hurricane Sandy Fact Sheet 1 – Cleaning Flooded Buildings
- o Hurricane Sandy Fact Sheet 2 – Foundation Requirements and recommendations for Elevated Homes
- o Hurricane Sandy Fact Sheet 3 – Building Science Support and Code Changes Aiding Sandy Recovery

- Developed in-person and online distance learning courses for FEMA's Emergency Management Institute, including an update to the four-day in-person course titled EMI E312, Fundamentals of Building Science, which reviews information pertaining to impacts of wind, flood, earthquake, and wildland-urban interface fire on the constructed environment, and explains key performance and construction issues related to floods, wind, wildfires, and earthquakes.
- Updated FEMA P-784, Substantial Damage Estimator (SDE) Version 2.1, after piloting Version 2.0 in Louisiana following Hurricane Isaac. SDE is a tool developed to assist state and local officials in determining substantial damage for residential and non-residential structures in accordance with a local floodplain management ordinance meeting the requirements of the National Flood Insurance Program. The tool can be used to assess flood, wind, wildfire, seismic, and other forms of damage.
- Developed a new fact sheet titled "Residential Tornado Safe Room Doors" (September 2014) that provides technical information on the selection and installation of safe room doors.
- Developed a new publication, FEMA P-1019, Emergency Power Systems for Critical Facilities: A Best Practices Approach to Improving Reliability (September 2014). This publication discusses the effects of natural hazards on electrical transmission and distribution infrastructure and on building systems. It discusses how to determine what facilities are critical, what equipment within a critical facility is needed to allow the facility to function and provide services, and how to provide emergency power to a critical facility.
- During the fiscal reporting period, FEMA has been updating FEMA P-320 and FEMA P361, due out in December 2014 and March 2015, respectively, in the next fiscal period.

FHWA: Training and Technology Transfer. Technology transfer of high performance computational analysis techniques is an important part of ANL's work to support and advance wind-related engineering and research programs at TFHRC. The technology transfer is accomplished by publishing techniques developed and the work done in reports and papers, presentations at conferences, and training courses in CFD capabilities and techniques offered by TRACC. A fourth two-day, hands-on training course was developed and provided to students, professionals, and U.S. Government representatives in March 2013 at ANL. The course consisted of lectures, practical examples, and detailed hands-on tutorials to demonstrate cutting edge capabilities and techniques using CFD. In addition, a fifth course with updated and expanded materials was developed and presented in March 2014.

FHWA: Aerodynamics Webinar Series. In November 2013, the FHWA launched a new webinar series on the aerodynamics of highway structures. The objectives of this new series are to introduce our stakeholders to the aerodynamics of highway structures, provide a review of highway aerodynamic issues, highlight updates and advancements of highway wind design specifications and guidelines, demonstrate advancements in aerodynamic analysis of highway structures, and provide an introduction to available mitigation methods. During Fiscal Years 2013 and 2014, a total of five one-hour-long sessions were developed and delivered by experts in the field, including FHWA staff, and at least four more sessions have been planned. The titles of the completed sessions are as follows: *FHWA Aerodynamics Program—from galloping to cruising*; *Introduction to wind hazards in highway engineering*; *Wind loads and aerodynamic design of bridges*; *Wind induced vibration of bridge cables*; and *Experimental methods for wind design.*

CONCLUSION

The destructive impacts of recent extreme windstorms – such as the historic Hurricane/PostTropical Cyclone Sandy in October 2012 and the violent tornadoes impacting Newcastle, Moore, and El Reno, Oklahoma, in May 2013 and Mayflower and Vilonia, Arkansas, in April 2014 – underscore the need for continued investments to make the buildings, cities, towns, and communities in which we live, and the infrastructure upon which we depend, not only less vulnerable to the forces of nature but also more resilient when disasters unfold. How these events affect us reflects not only the power of nature, but also the decisions we make in how we build and safeguard our communities.

Fundamental research on the meteorological aspects of wind hazards is continuing with support from the Federal agencies, and there are a number of areas where additional knowledge and action have reduced the impact of windstorms to lives and property. Notable successes have been demonstrated in the following areas:

- Advances in NOAA's satellite-based observations, supercomputers, and data assimilation and modeling have reduced average hurricane forecast track errors significantly, to about half of what they were 15 years ago;
- Advances in the use of aircraft data have demonstrated the potential for significant improvements in hurricane intensity forecasts (20 to 40 percent), breaking a 30-year logjam in intensity forecast improvements;
- Improvements in the understanding of tornadogenesis as well as forecasting and prediction have enabled NWS to double the average lead time for tornado warnings over the past two decades to 13 minutes;

- FEMA publications presenting design and construction guidance for safe rooms have been available since 1998. Since that time, over one million copies have been distributed and thousands of safe rooms have been built, and a growing number of these safe rooms have already saved lives in actual events. There has not been a single reported failure of a safe room constructed to FEMA criteria;
- Gains in structural engineering research and mitigation assessment have allowed FEMA to update its published guidance for better and more efficient safe room design and construction against windstorms, much of which has also been incorporated into the ICC 500 Standard for the Design and Construction of Storm Shelters;
- FEMA successfully proposed two building code changes for the 2015 IBC, which require ICC 500-compliant storm shelters in new schools and first-responder facilities in the areas of the Nation with the highest tornado risk;
- Results of wind engineering research and post-storm studies by NIST have led to changes in the wind loading provisions of the ASCE 7 Standard and International Building Code that enhance life safety and property protection for windstorms;
- Improvements in tornado intensity estimation (developed jointly by NIST, NOAA, and Texas Tech University) have allowed the Enhanced Fujita (EF) tornado scale to be introduced by the NWS in 2007 to reflect more accurate examinations of tornado damage surveys and to align wind speeds more closely and effectively with associated storm damage;
- The results of research funded by NSF grants have strengthened windstorm disaster evacuation management practices in the United States through advancements in planning and preparedness efforts and a better understanding of how communities perform in hosting incoming evacuees; and
- Results of wind engineering research by FHWA have contributed to a better understanding of bridge cable aerodynamics and effectiveness of associated wind mitigation techniques, improved techniques for physical and computational modeling of wind hazards to transportation structures, as well as updates of design guides and specifications.

There are opportunities for further improvements in windstorm prediction capabilities and for additional observations to advance understanding of how the built environment responds to windstorm events. In addition, efficient warning systems and social analysis can improve warning response, mitigate user complacency, better support effective decision making, and reduce the losses associated with devastating windstorms.

To help address these and related issues, the NWIRP agencies, academia, and the private sector continue to build on past activities, including those identified in this report, to identify and prioritize specific measurement science research and development needs for windstorm impact reduction.

APPENDIX A: KEY ABBREVIATIONS

ANL	Argonne National Laboratory
AOML	Atlantic Oceanographic and Meteorological Laboratory
AASHTO	American Association of State Highway and Transportation Officials
ASCE	American Society of Civil Engineers
BFE	Base Flood Elevation
CFD	Computational Fluid Dynamics
CISS	Critical Infrastructure-based Societal Systems
CSM	Computational Structural Mechanics
DOE	Department of Education
EF	Enhanced Fujita
ESRL/GSD	Earth System Research Laboratory Global Systems Division
FEMA	Federal Emergency Management Agency
FHWA	Federal Highway Administration
FLASH	Federal Alliance for Safe Homes
FPHLM	Florida Public Hurricane Loss Model
GOES-R	Geostationary Operational Environmental Satellite – R Series
HDPE	High Density Polyethylene Pipe
HFIP	Hurricane Forecast Improvement Program
HRRR	High Resolution Rapid Refresh
HWT	Hazardous Weather Testbed
IBC	International Building Code
IFEX	Intensity Forecasting Experiment
IRC	International Residential Code
JHT	Joint Hurricane Testbed
MAT	Mitigation Assessment Team
MPAR	Multifunction Phased Array Radar
NASA	National Aeronautics and Space Administration
NCHRP	National Cooperative Highway Research Program
NCST	National Construction Safety Team
NHC	National Hurricane Center
NIST	National Institute of Standards and Technology
NOAA	National Oceanic and Atmospheric Administration
NRCC	National Research Council of Canada
NSF	National Science Foundation
NSSL	National Severe Storms Laboratory
NSTC	National Science and Technology Council
NWIRA	National Windstorm Impact Reduction Act
NWIRP	National Windstorm Impact Reduction Program
NWS	National Weather Service
ONR	Office of Naval Research
OSTP	Office of Science and Technology Policy
OUN	National Weather Service Forecast Office in Norman, Oklahoma
RAP	Rapid Refresh Model

SDE	Substantial Damage Estimator
SDR	Subcommittee on Disaster Reduction
SPC	Storm Prediction Center
TC	Tropical Cyclone
TFHRC	Turner Fairbank Highway Research Center
TRACC	Transportation Research and Analysis Computing Center
VORTEX2	Verification of the Origins of Rotation in Tornadoes Experiment 2
WoF	Warn-on-Forecast
WRN	Weather-Ready Nation

End Notes

1 HUD, Management's Discussion and Analysis Hurricane Sandy, https://portal.hud.gov/hudportal/documents/huddoc?id=afrfy13_sandy.pdf.

2 NOAA/NWS Service Assessment, Hurricane/Post-Tropical Cyclone Sandy, October 22-29, 2012, http://www.nws.noaa.gov/os/assessments/pdfs/Sandy13.pdf.

3 Ibid.

4 NOAA/NWS fatality and injury data (2013) for tornado, thunderstorm wind, hurricane/tropical storm, and high wind hazards, http://www.nws.noaa.gov/om/hazstats/sum13.pdf.5 NOAA/NWS property loss data (2013) for tornado, thunderstorm wind, hurricane/tropical storm, and high wind hazards, http://www. nws.noaa.gov/om/hazstats/sum13.pdf

5 NOAA/NWS property loss data (2013) for tornado, thunderstorm wind, hurricane/tropical storm, and high wind hazards, http://www.nws.noaa.gov/om/hazstats/sum13.pdf

6 NWS Weather Forecast Office, Norman, OK, The Tornado Outbreak of May 20, 2013, http://www.srh.noaa.gov/oun/?n=events-20130520.

7 NOAA/NWS Service Assessment, May 2013 Oklahoma Tornadoes and Flash Flooding, http://www.nws.noaa.gov/os/assessments/pdfs/13oklahoma_tornadoes.pdf.

8 NWS Weather Forecast Office, Norman, OK, May 31-June 1, 2013 Tornado and Flash Flood Event, http://www.srh.noaa.gov/oun/?n=events-20130531-elreno.

9 NWS Weather Forecast Office, Little Rock, AR, Tornadoes on April 27, 2014 (Ground Photos), http://www.srh.noaa.gov/lzk/?n=svr0414cpics.htm.

10 NOAA National Climatic Data Center, Storm Events Database, http://www.ncdc.noaa.gov/ stormevents/ event details.jsp?id=504760.

11 NOAA/NWS Storm Prediction Center, Monthly and Annual U.S. Tornado Summaries, http://www.spc.noaa.gov/climo/online/monthly/newm.html.

12 http://www.sdr.gov/

13 NOAA National Hurricane Center, NHC Tropical Cyclone Forecast Verification, http://www.nhc.noaa.gov/verification/verify5.shtml.

14 NOAA National Hurricane Center, presentation by Richard J. Pasch, Hurricane Intensity Forecast Improvement: Is It Possible?, April 21, 2011, http://www.nhc.noaa.gov/outreach/presentations/2011_ Intensity Forecasting Improvement_Pasch.pdf.

15 NOAA, Tornadoes 101: Stay alert and stay alive, http://www.noaa.gov/features/ 03_protecting/tornadoes 101c.html.

16 From VORTEX2 website: http://www.vortex2.org/home/.

17 NIST Advances in Computer-Aided and Computational Methods in Wind Engineering, by D. Yeo, F. Lombardo, D. Banerjee, E. Letvin, F. Potra, E. Simiu, and M. Levitan, Proceedings of the ATC/SEI Advances in Hurricane Engineering Conference, October 2012.

18 Practical CFD Simulations of Wind Tunnel Tests, by D. Banerjee, S. Hemley, R. McDermott, D. Yeo, F. Lombardo, and M. Levitan. Proceedings of the 12th Americas Conference on Wind Engineering, June 2013.

19 Hurricane Isaac in Louisiana: Building Performance Observations, Recommendations, and Technical Guidance, Mitigation Assessment Team Report (P-938), Federal Emergency Management Agency, March 2013.

Available at http://www.fema.gov/media-library-data/20130726-1908-25045-0581/fema_p_ 938_isaac_mat.pdf.

[20] Hurricane Sandy in New Jersey and New York: Building Performance Observations, Recommendations, and Technical Guidance, Mitigation Assessment Team Report (P-942), Federal Emergency Management Agency, November 2013. Available at http://www.fema.gov/media-library-data/1386850803857-025eb299df32c6782fdcbb6f69b35b13/Combined_Sandy_MAT_Report_508post.pdf.

[21] Tornado: Moore, Oklahoma, May 20, 2013: Safe Room Performance, Observations, and Conclusions, Formal Observation Report (P-1020), Federal Emergency Management Agency, August 2014. Available at http://www.fema.gov/media-library-data/1418307179636-a018b8744801f8a770e6ea5c65eb5a4d/FEMA_P-1020_Moore_Tornado_Report_508.pdf.

[22] Preliminary Reconnaissance of the May 20, 2013, Newcastle-Moore Tornado in Oklahoma. By E. Kuligowski, L. Phan, M. Levitan, and D. Jorgensen. National Institute of Standards and Technology Special Publication 1164, December, 2013. Available at http://www.nist.gov/manuscript-publication-search.cfm?pub_id=914721.

[23] Final Report - National Institute of Standards and Technology (NIST) Technical Investigation of the May 22, 2011, Tornado in Joplin Missouri. By E. Kuligowski, F. Lombardo, L. Phan, M. Levitan, and D. Jorgensen. NCSTAR 3, March 2014. Available at http://www.nist.gov/manuscript-publication-search.cfm?pub_ id= 915628.

[24] Ibid.

[25] Proceedings of the 29th U.S. – Japan Bridge Engineering Workshop, Public Works Research Institute, Tsukuba Science City, Japan, November, 2013. Available at http://www.pwri.go.jp/eng/ujnr/ujnr.htm.

[26] Measurement Science R&D Roadmap for Windstorm and Coastal Inundation Impact Reduction (NIST GCR 14 973-13), National Institute of Standards and Technology, 2014. Available at http://www.nist.gov/manuscript-publication-search.cfm?pub_id=915541.

[27] Dynamic Properties of Stay Cables on the Penobscot Narrows Bridge, by H.R. Bosch and J.R. Pagenkopf, Report No. FHWA-HRT-14-067, Federal Highway Administration, McLean, VA, September 2014. Available at http://www.fhwa.dot.gov/publications/research/infrastructure/structures/bridge/14067/index.cfm.

[28] Dynamic Properties of Stay Cables on the Bill Emerson Bridge, by H.R. Bosch and J.R. Pagenkopf, Proceedings of the 12th Americas Conference on Wind Engineering, Seattle, Washington, June 2013.

[29] Wind Tunnel Investigations of an Inclined Stay Cable with a Helical Fillet, by G.L. Larose and A. D'Auteuil, Report No. FHWA-HRT-14-070, Federal Highway Administration, McLean, VA, September 2014. Available at http://www.fhwa.dot.gov/publications/research/infrastructure/structures/bridge/14070/index.cfm.

[30] Wind-Induced Vibrations of Dry Inclined Stay Cables in the Critical Reynolds Number Range, by J.B. Jakobsen, G.L. Larose, A. D'Auteuil, J.H.G. Macdonald, and H.R. Bosch, Proceedings of the 6th Symposium on Strait Crossings, Bergen, Norway, June 2013.

[31] Wind-Tunnel Investigations of an Inclined Stay Cable with a Helical Fillet, by G.L. Larose, A. D'Auteuil, H.R. Bosch, J.B. Jakobsen, and J.H.G. Macdonald, Proceedings of the 6th European-African Conference on Wind Engineering, Cambridge, UK, July 2013.

[32] Sectional Load Characteristics of a Dry Inclined Helical Filleted Cable, by H. Christiansen, J.B. Jakobsen, J.H.G. Macdonald, G.L. Larose, and H.R. Bosch, Proceedings of the 13th Conference of the Italian Association for Wind Engineering, Genova, Italy, June 2014.

[33] Comparison of the Aerodynamics of a Bridge Cable with Helical Fillets in Smooth and Turbulent Flow, by H. Christiansen, J.B. Jakobsen, G.L. Larose, J.H.G. Macdonald, A. D'Auteuil, and H.R. Bosch, Proceedings of the Symposium on the Dynamics and Aerodynamics of Cables, Copenhagen, Denmark, September 2014.

[34] Wind Tunnel Model of the Turner-Fairbank Highway Research Center Aerodynamics Laboratory, by S.A. Lottes and C. Bojanowski, Report No. ANL/ESD/13-3, Argonne National Laboratory, Argonne, IL, June 2013. Available at http://www.osti.gov/scitech/biblio/1111295.

[35] Computer Modeling and Analysis of Truck Generated Salt-Spray Transport Near Bridges, by S.A. Lottes and C. Bojanowski, Report No. ANL/ESD/13-1, Argonne National Laboratory, Argonne, IL, July 2013. Available at http://www.osti.gov/scitech/biblio/1087817.

[36] Mitigation of Wind-Induced Vibration of Stay Cables: Numerical Simulations and Evaluations, by S. Park and H.R. Bosch, Report No. FHWA-HRT-14-049, Federal Highway Administration, McLean, VA, August 2014. Available at http://www.fhwa.dot.gov/publications/research/infrastructure/structures/bridge/14049/index.cfm.

In: National Windstorm Impact Reduction Program
Editor: Ralph Simmons
ISBN: 978-1-63484-291-4

Chapter 2

WINDSTORM IMPACT REDUCTION IMPLEMENTATION PLAN*

National Science and Technology Council

1. EXECUTIVE SUMMARY

The tragedy caused by Hurricanes Katrina and Rita in August and September 2005, the unprecedented hurricane season of 2004 in which five hurricanes made landfall in Florida, and the May 1999 outbreak of damaging tornados in Oklahoma underscore the significant and growing risks to our society due to wind hazards. Public Law 108-360, known as the National Windstorm Reduction Act of 2004, was signed into law by President Bush to reduce the risk wind hazards pose to life and property. The law increased national attention on wind hazard reduction efforts, which will require significantly improved cooperation and coordination between Federal agencies, improved coordination with states and local governments and increased, focused Federal investment to reduce wind hazards.

Although there are current and ongoing activities related to, or focused on, wind hazards it is clear that these efforts are more often merely small parts of larger efforts and are not coordinated among the agencies. For example, the National Oceanic and Atmospheric Administration (NOAA) makes weather predictions regarding several physical parameters of which wind is only one part; the National Institute of Standards and Technology (NIST) studies structural hazards ranging from earthquakes to fires, and winds are only one component; and the Federal Emergency Management Agency (FEMA) conducts evacuation planning studies, promotes wind preparedness activities, and advocates enhancements to the nation's model building codes as part of multi-hazard programs.

Each agency's effort toward wind hazard reduction is detailed in Appendix C. There appears to be virtually no duplication of effort among the agencies but there are gaps in knowledge, implementation and coordination.

* This is an edited, reformatted and augmented version of a document issued by the Office of Science and Technology Policy, April 2006.

In accordance with the legislation, a coordinated Federal effort, in cooperation with other levels of government, academia, and the private sector, will improve the understanding of windstorms and their impact, and develop and encourage implementation of cost-effective mitigation measures to reduce those impacts while promoting community resilience. We recommend a coordinated, comprehensive multi-agency, multi-disciplinary group be established as a working group of the National Science and Technology Council's Committee on Environment and Natural Resources Subcommittee on Disaster Reduction to reduce the impact of wind hazards by facilitating better communication among agencies, effectively allocating collective resources and operating within a common framework. This working group shall meet at least quarterly, report to the Subcommittee on Disaster Reduction annually and work with state, local officials and non-government organizations as appropriate. All Federal agencies contributing to this document shall be members of the working group and the chair of the working group will rotate between NIST, NSF, NOAA and FEMA with each Agency serving a two-year term as chair.

A coordinated portfolio that builds on current efforts in research and mitigation activities should be developed including:

- Assessing individual and community capability to respond to wind events, including vulnerability analyses, risk perception, risk communication, risk management, communication of wind warnings and public response, evacuation capability, and public knowledge of appropriate protective actions for wind events, especially among vulnerable populations
- Evaluating the response of the built environment and critical infrastructure to wind events by investigating aerodynamic response, load path, ultimate capacity and the performance of the building envelope
- Assessing the impact of wind and windborne debris or wind and water/ice/snow
- Examining the interaction between wind and storm surge to determine the impact on building foundations and critical infrastructure
- Exploring the near-ground and channeling/shielding effects of winds on buildings through testing and instrumentation
- Developing new technologies and ground, airborne and satellite based observing systems to improve knowledge and understanding of windstorms and the wind variability within those storms
- Measuring the response of bridges and other highway structures to wind events, including stability, serviceability and functionality leading up to and through extreme events
- Developing and implementing technologies for rapid repair and restoration of critical infrastructure and critical services

There could be improvements in and enhancement of:

- Local, state, regional and federal coordination in response to wind hazard events
- Evacuations through more informed planning and annual drills
- Local and regional preparedness through public-private partnerships fostering outreach and technology transfer programs

- Windstorm impact reduction practices through training and outreach programs that enhance state and local capabilities

2. Purpose and Scope

Public Law 108-360, Title II, known as the National Windstorm Impact Reduction Act of 2004, was signed into law by President Bush on October 25, 2004. This law seeks to measurably reduce the loss of life and property from windstorms. The law states that,

> No later than 90 days after the date of enactment of this Act, the Director of the President's Office of Science and Technology Policy shall establish an Interagency Working Group consisting of representatives of the National Science Foundation (NSF), the National Oceanic and Atmospheric Administration (NOAA), the National Institute of Standards and Technology (NIST), the Federal Emergency Management Agency (FEMA), and other Federal agencies as appropriate. The Interagency Working Group will be responsible for the planning, management, and coordination of the Program, including budget coordination.

- NIST shall support research and development to improve building codes and standards and practices for design and construction of buildings, structures, and lifelines.
- NSF shall support research in engineering and the atmospheric sciences to improve the understanding of the behavior of windstorms and their impact on buildings, structures and lifelines.
- NOAA shall support atmospheric sciences research to improve the understanding of the behavior of windstorms and their impact on buildings, structures, and lifelines.
- FEMA shall support the development of risk assessment tools and the effective mitigation techniques, windstorm-related data collection and analysis, public outreach, information dissemination, and implementation of mitigation measures consistent with the Agency's all-hazards approach.

The Act further stipulates that not later than one year after the date of enactment of this title, the Interagency Working Group shall develop and transmit to Congress an implementation plan for achieving the objectives of this Program. This document is that plan.

3. Introduction

3.1. Costs of Structural Wind Damage

The 2005 hurricane season brought storms that wreaked havoc, causing large-scale damage from wind and water across many states. It was the most costly hurricane season in U.S. history. Hurricanes Katrina, Rita and Wilma demonstrated in dramatic fashion how costly these events can be, both in lives lost and property destroyed. The costs of wind damage to the built environment from wind hazards continue to increase, even when adjusted for inflation. These devastating events were only the most recent examples of the worldwide impact of windstorms.

According to a report published by RAND[1], windstorms caused almost two-thirds of the \$145B total 2004 uninsured losses. The State of Florida was struck by four hurricanes in 2004, and uninsured losses there alone reached \$42B. Losses due to windstorms include damages from wind, storm surge and flooding. The RAND report indicates that, in the United States, losses due to hurricanes and tornadoes total \$6B per year.

Although these losses show an increasing trend over the last decade, improvements in forecasts, preparation and mitigation have significantly reduced the number of lives lost over the last century. Deaths and injuries from wind-hazard extreme events, as estimated before Hurricane Katrina, average about 100 deaths and 1,250 injuries per year. Social vulnerability analysis has shown that demographic subpopulations, including those of lower socioeconomic status, are at higher levels of risk from wind events.

3.2. Previous Wind Impact Reduction Efforts

Improved wind forecasting, wind characterization, wind engineering and design, improved building standards and codes, and mitigation activities, have led to a better understanding of wind hazards, and helped reduce their effects on the built environment.

Social sciences, research for the past fifty years has focused upon producing more effective hurricane and tornado warnings and protective actions on the part of local officials and individuals. Research has led to significant strides in understanding the warning process and implementing effective risk communication. Improved hurricane evacuation planning is the direct result of research over the past thirty years. Thousands of lives may have been saved by improvements in warning and evacuation to date. But, given the devastating impacts of the 2005 hurricane season, clearly more improvements must be made.

Research into individual protective actions also has improved public response to tornadoes. Decades of research on community emergency preparedness and response have produced more effective state and local capabilities. It is an ongoing process to maintain effective capabilities at the state and local level.

Key Federal agency players include:

- Federal Emergency Management Agency (FEMA)
- National Oceanic and Atmospheric Administration (NOAA), Department of Commerce
- National Science Foundation (NSF)
- National Institute of Standards and Technology (NIST), Department of Commerce
- Federal Highway Administration (FHWA)
- Housing and Urban Development (HUD)
- National Aeronautics and Space Administration (NASA)
- U.S. Army Corps of Engineers (USACE)

Also playing a critical role are:

- State governments
- County governments

- City governments
- State-supported public universities

Finally, non-government organizations also continue to play significant roles:

- American Association of Wind Engineers (AAWE)
- American Society of Civil Engineers (ASCE)
- Institute for Building and Home Safety (IBHS)
- Private universities

3.3. Recent Preparedness and Mitigation Actions

Hurricane Katrina set records for a single disaster event and early indications are that preparedness and mitigation prior to this event were not adequate. The State of Florida had a record hurricane year in 2004 emphasizing the fact that, while some progress has been made in mitigation measures to protect lives and property, much remains to be done. A recent survey conducted by IBHS found that more than half of Florida's homeowners have taken action to protect their homes from hurricanes and an even larger number recognize the importance of the state's building codes in reducing property losses from these events. Many think the codes should be even stronger. States, such as Texas, are adopting hurricane coastline wind load building codes and these are being implemented at the county and city levels. NOAA All-Hazards Radio and the National Weather Service's StormReady program are encouraging preparedness. Advances in wind engineering and design are increasing the effectiveness and popularity of affordable shelters, such as in-residence tornado shelters, based on design and construction guidance developed by FEMA.

4. Current Activities and Capabilities

Several Federal agencies have missions and mission capabilities that are essential to achieving the objectives of the National Windstorm Act. (See Appendix C for Federal agency-specific contributions.) It is difficult to estimate the amount of money that Federal agencies invest in wind hazard related activities alone because of the broad nature of agency missions. Understanding wind events, assessing wind damage, reducing the impacts and enhancing community resilience are often part of larger efforts to understand severe weather, hurricanes, building safer structures, etc. These larger efforts have myriad purposes, of which wind is often a tiny piece, and almost never the primary motivator. Therefore, the total investment by Federal agencies that can be explicitly and solely applied to wind hazard reduction is quite small in relation to the magnitude of losses and the national impact of those losses.

4.1. Understanding, Predicting and Forecasting Windstorm Hazards

Federal agency efforts to understand, predict, design for, and forecast windstorm hazards range from basic research funded by NSF and operational forecasts provided by NOAA to evacuation studies jointly funded by FEMA and USACE. NIST and FHWA focus on research aimed at understanding, predicting and designing for wind effects on structures, critical infrastructure, and creating the knowledge needed to develop improved design code provisions. NASA's work provides satellite based observational data and develops new observing systems that help underpin and improve wind predictions and forecasts. FEMA provides State and local governments the tools and products to enhance their capability to effectively manage loss reduction programs.

The focus on understanding and predicting of windstorm hazards and risks within any one of these agencies is minimal at this time. Wind is only one relatively small component of a broader suite of hazards addressed by each agency. For example, the observations of NOAA and NASA and the forecasts from NOAA address temperature, precipitation, moisture, air pressure and wind. However, the wind measurements taken are seldom at the small space and time scales required by engineers. In addressing hazards, FEMA, NIST and FHWA address issues related to buildings and public infrastructure for mitigating earthquakes, floods, fires, wind and other hazards making wind just one small component of their overall responsibility.

Nevertheless, efforts by these agencies have led to some improvements in the understanding and prediction of the winds themselves and in the development of programs for mitigating their effects. The spatial and temporal scale of wind predictions are decreasing and becoming more accurate. There has been a gradual increase in the understanding of structural response to winds, but progress has been uneven, particularly in the area of high winds combined with storm surge, flying debris, and ice loading on structures, among others. Another area where more attention is needed is the transfer of current research and knowledge into effective guidance and practice, including cost-effective mitigation.

4.2. Assessing the Impact of Wind Hazards

The investigation of wind-induced damage to buildings and critical infrastructure should happen immediately following a wind-hazard event. These assessments of wind impacts are essential to providing new knowledge about how wind behaves at varying elevations and around differing structural envelopes. To assess impact of wind hazard loadings on structures and to gain better understanding of the impact of winds and windstorm on their performance, it is essential to gather data through instrumentation before and field data collection after the hazard event. Through these types of efforts new or enhanced test procedures, tools for predicting structural response to wind and models for computer simulation of wind/structure interaction and loss estimation and improvements in design can become available. Damage assessment is also a focus of social science research as well as household, business and community recovery, but is not necessarily wind related.

4.3. Reducing the Impact of Wind Hazards

Successfully reducing the impact of wind hazards requires that actions be taken, directly or indirectly, to change or enhance existing building practices, infrastructure resilience, social behavior patterns and evacuation processes. Improvements in warning systems, evacuation planning and building technology have reduced the threat of windstorms to people even while the total number of people, buildings and critical infrastructure exposed to windstorms has grown dramatically. The result is that even while the threat of injury or death is being reduced, the total amount of damage and loss continues to rise. Many improvements have been implemented but much more needs to be done. Techniques have been developed to estimate wind effects that account realistically for wind directionality characteristics. Further, estimation methods have been developed to help assure higher safety levels for tall buildings that experience dynamic effects. The continuing improvement of the Nation's model building codes and standards with respect to improved design and construction provisions for wind resistance have had a significant impact on building and infrastructure performance. Ongoing efforts to develop performance based approaches to building design, especially critical and essential facilities, promise to bring further improvements. The additional development and use of innovative computer-intensive, user friendly methods to quantify wind loading can reduce errors in the estimation of wind loads by as much as fifty percent and can result in stronger structures built for lower costs. This is of interest to the private sector, state and local government, as well as Federal agencies, including FEMA and USAGE, which are focused on increasing building survivability through optimized designs, building in hostile environments, and in hurricanes and other wind storms.

New designs that are evaluated in wind tunnels to ensure safety and performance are reducing the impact of winds and windstorms on highway structures. Efforts also have been made to promote structural health monitoring which permits evaluation of changes within a structure over time as a way to determine degradation before a catastrophic event and to incorporate monitoring instrumentation into major new structures.

4.4. Preparedness and Enhancing Community Resilience

Preparedness is the advance capacity to respond to the consequences of a hazard event. This means having emergency plans in place concerning what to do and where to go if a warning is received or a hazard is observed. FEMA, in partnership with NOAA and USAGE, develops the tools and products to develop effective State and local plans for evacuation and sheltering. NSF undertakes basic research on warning, evacuation, emergency planning and response, and vulnerability analysis.

Gommunity resilience, especially in vulnerable populations, can be enhanced through a variety of means including outreach and awareness programs, partnerships with Federal and State agencies to improve the resilience of buildings and infrastructure, increased preparedness exercises, effective mitigation planning efforts, and through improvements in local building code adoption and enforcement activities.

Numerous informational and educational materials exist on protection of individuals and property in high wind events, including hurricanes, tornadoes and straight line winds from thunderstorms. This information is repeated in brief during severe weather warnings. Coastal regions impacted by wind are encouraged to be prepared, have mitigation plans in place and establish sound building practices.

Within NOAA's National Weather Service, an outreach and education program called StormReady (http://www.stormready.noaa.gov/) has now reached communities in all 50 States. StormReady helps arm America's communities with the communication and safety skills needed to save lives and property– before and during the event. Working with local emergency managers, the Weather Service helps communities strengthen local safety programs. All potential weather risks to a community are included. Wind is usually a significant risk to life and property in most locations.

In addition, social science research that includes social vulnerability analysis as well as organizational and community emergency preparedness and response has been sponsored by NSF.

5. Gaps

5.1. Understanding, Predicting and Forecasting

There are a number of areas where additional knowledge and action may reduce the impact of windstorms. Among these are enhanced full-scale wind measurement and structure performance with more sites and broader distribution, better and more robust instruments and networking, improved coordination between agencies and jurisdictions and setting standards for sensors, deployment, measurements and data storage.

Fundamental research on the meteorological aspects of wind hazards is continuing under support from NSF and other agencies. The interagency U.S. Weather Research Program has placed a high priority on improved knowledge of hurricane dynamics and developing improved forecasts of hurricane intensity. Major gaps in this research arena include the need for better understanding of the influence on intensity of: ocean heat content; environmental wind and thermodynamic structures and internal hurricane dynamics, such as the impact of hurricane eye wall and rain band interactions.

Improved prediction of wind storms is essential, but almost totally lacking are observations and understanding of the small scale wind structure in time and space to which the built environment responds within a larger windstorm event. Simulation of such detailed wind structure in wind tunnels is useful, but observations in the real environment near full-scale buildings must be obtained. These observations will provide necessary information on the interaction between individual structures and various types of complex built environments on the one hand and high winds on the other.

Methods and tools for wind hazard exposure predictions are required for structural design purposes. Improved methodologies for site-specific wind climate models and more refined and locally-detailed wind speed/hazard maps are therefore a critical need.

5.2. Assessing the Impact of Wind Hazards

Tools for assessing the impact of wind hazard events should be developed and improved. A greater effort must be made to study and learn from the aftermath of wind events and investigations of wind-impacted structures should be enhanced by including a broader spectrum of structures, including critical infrastructure. More comprehensive data on wind and windstorms should be collected and data exchange on damage and loss should be encouraged. New methods to predict the risk or loss and damage due to windstorms should be developed with appropriate simulation and modeling tools. Improved understanding of the effects of wind-borne debris on structures as well as the additional risk to structures from wind-driven rain, ice and hail are other important issues. For instance, the wind (speed and turbulent structure) and icing thresholds of danger to these buildings and infrastructure is not well understood.

Social scientists should examine the impact of wind hazards on individuals, businesses and communities, including vulnerable populations, by leveraging National Science Foundation funding to universities and non-government organizations.

Improved methods for assessing social and economic costs are also needed. Inclusion of detailed loss data from the insurance industry in this assessment effort is of vital importance.

5.3. Reducing the Impact of Wind Hazards

New, more accurate methods of understanding and assessing risk perception, risk communication, risk management, and designing for wind will reduce the impact of wind hazard events. Systematic establishment of design and retrofitting requirements should continue and a closer partnership between the design, construction and industry communities should be established to allow exploration, development and application of innovative technologies. Another area of emphasis should address ongoing programs to ensure that State and local knowledge and capabilities regarding wind hazards are high. Improvement in building code provisions and in wind loading provisions such as those contained in the ASCE 7 Standard is a central part of the impact reduction effort. Improved enforcement of new and existing building code requirements in areas of high risk to wind-borne hazards is equally important. Improved enforcement and inspection by local building officials is a very effective way to decrease the potential for damage from improper construction. Attention to damage generated by wind driven missiles and debris on building envelopes and glazing systems is needed.

5.4. Preparedness and Enhancing Community Resilience

Information regarding risk and preparedness should be broadly distributed in a coordinated fashion. Gaps include the lack of sufficiently effective decision making tools for warning and evacuation; increased understanding of household and community adoption of preparedness measures; improved understanding of the role of improvisation and resilience in emergency preparedness and response; increased understanding of community physical, economic and social recovery from wind related disasters.

Additional work is needed to understand how individuals, especially those in vulnerable populations, perceive their risk to wind hazards and how they might best receive warning messages. This knowledge will help individuals understand their likelihood of preparedness and the potential need to target messages to specific audiences and geographical locations.

As with any other kind of disaster, community resilience is further determined by how quickly essential services are restored after a wind event. Damage resistant infrastructure must be designed and new technical methods must be developed for rapid repair of damaged infrastructure and restoration of services. Without restoration of essential services, like electricity and gas for cooking, potable water for drinking or telephone lines for communication, extreme community disruption will occur.

6. RECOMMENDATIONS

In accordance with the legislation, a coordinated Federal effort, in cooperation with other levels of government, academia, and the private sector, will improve the understanding of windstorms and their impact, and develop and encourage implementation of cost-effective mitigation measures to reduce those impacts while promoting community resilience. We recommend a coordinated, comprehensive multi-agency, multi-disciplinary working group be established as a working group of the National Science and Technology Council's Committee on Environment and Natural Resources Subcommittee on Disaster Reduction to reduce the impact of wind hazards by facilitating better communication between agencies, effectively allocating collective resources and operating within a common framework. This working group shall meet at least quarterly, report to the Subcommittee on Disaster Reduction annually and work with state, local officials and non-government organizations as appropriate. All Federal agencies contributing to this document shall be members of the working group and the chair of the working group will rotate between NIST, NSF, NOAA and FEMA with each Agency serving a two-year term as chair.

A coordinated portfolio of research and other activities should build on existing efforts and should include:

- Assessing individual and community capability to respond to wind events, including vulnerability analyses, risk perception, risk communication, risk management, communication of wind warnings and public response, evacuation capability, and public knowledge of appropriate protective actions for wind events, especially among vulnerable populations
- Evaluating the response of the built environment and critical infrastructure to wind events by investigating aerodynamic response, load path, ultimate capacity and the performance of the building envelope
- Assessing the impact of wind and windborne debris or wind and water/ice/snow
- Examining the interaction between wind and storm surge to determine the impact on building foundations and critical infrastructure
- Exploring the near-ground and channeling/shielding effects of winds on buildings through testing and instrumentation

- Developing new technologies and ground, airborne and satellite based observing systems to improve knowledge and understanding of windstorms and the wind variability within those storms
- Measuring the response of bridges and other highway structures to wind events, including stability, serviceability and functionality leading up to and through extreme events
- Developing and implementing technologies for rapid repair and restoration of critical infrastructure and critical services

There could be improvements in and enhancement of:

- Windstorm prediction
- Local, state, regional and federal coordinated response capabilities following wind hazard events, including field validation and data collection capabilities for buildings, critical infrastructure and essential facilities
- Windstorm damage and loss estimation modeling tools
- Standards and technologies that will enable cost-effective, state-of-the-art windstorm-resistant provisions to be adopted as part of state and local building codes

These improvements and enhancements would enable more effective:

- Local, state, regional and federal coordination in response to wind hazard events
- Evacuations through more informed planning and annual drills
- Local and regional preparedness through public-private partnerships fostering outreach and technology transfer programs
- Windstorm impact reduction practices through training and outreach programs that enhance state and local capabilities

APPENDIX A: ELEMENTS OF DISASTER RISK REDUCTION AND HAZARD MANAGEMENT

The hazard research and management community employs a range of terminology to describe its activities, but no definitive, comprehensive list of these terms and their definitions exists. However, hazard risk reduction and disaster management activities can be grouped largely under nine broad concepts: research and development, hazard identification, risk assessment, risk communication, prediction, mitigation, preparedness, response, and recovery. Each of these rubrics includes critical science and technology elements, and, taken together, they form the nation's toolbox for reducing vulnerability to disaster risk.

1. *Disaster Process Research and Development (R&D)*—the science activities dedicated to improving understanding of the underlying processes and dynamics of each type of hazard. R&D includes fundamental and applied research on geologic, meteorological, epidemiological, and fire hazards; development and application of remote sensing technologies, software models, infrastructure models, organizational

and social behavior models; emergency medical techniques; and many other science disciplines applicable to all facets of disasters and disaster management.

2. *Hazard Identification*—determining which hazards threaten a given area. This includes understanding an area's history of hazard events and the range of severity of those events. The continuous study of the nation's active faults, seismic risks, and volcanoes are included in this category, as are efforts to understand the dynamics of hurricanes, tornadoes, floods, droughts, and other extreme weather events.
3. *Risk Assessment*—determining the impact of a hazard or hazard event on a given area. This includes advanced scientific modeling to estimate loss of life, threat to public health, structural damage, environmental damage, and economic disruption that could result from specific hazard event scenarios. Risk assessment takes place both before and during disaster events.
4. *Risk Communication*—public outreach, communication, and warning at every stage of hazard management. Risk communication includes raising public awareness and effecting behavioral change in the areas of mitigation and preparedness; the deployment of stable, reliable, and effective warning systems; and the development of effective messaging for inducing favorable community response to mitigation, preparedness, and warning communications.
5. *Mitigation*—sustained actions taken to reduce or eliminate the long-term risk to human life and property from hazards based on hazard identification and risk assessment. Examples of mitigation actions include planning and zoning to manage development in hazard zones, storm water management, fire fuel reduction, acquisition and relocation of flood-prone structures, seismic retrofit of bridges and buildings, installation of hurricane straps, construction of tornado safe rooms, and flood-proofing of commercial structures.
6. *Prediction*—predicting, detecting, and monitoring the onset of a hazard event. Federal agencies utilize weather forecast models, earthquake and volcano monitoring systems, remote sensing applications, and other scientific techniques and devices to gather as much information as possible about the what, when, and where of a potential hazard, as well as the severity of each threat.
7. *Preparedness*—the advance capacity to respond to the consequences of a hazard event. This means having plans in place concerning what to do and where to go if a warning is received or a hazard is observed. Communities, businesses, schools, public facilities, families, and individuals should have preparedness plans.
8. *Response*—the act of responding to a hazard event. Hazard response activities include evacuation, damage assessment, public health risk assessment, search and rescue, fire suppression, flood control, and emergency medical response. Each of these response activities relies heavily on information and communication technologies.
9. *Recovery*—activities designed to restore normalcy to the community in the aftermath of a hazard event. Recovery activities include restoring power lines, removing debris, draining floodwater, rebuilding, and providing economic assistance programs for disaster victims. As with response, the recovery process relies heavily on the availability of up-to-date data and information about the various community sectors, and on the technology to obtain and communicate that information.

APPENDIX B: REFERENCE

1. Meade, C. and Abbot, M., "Assessing Federal Research and Development for Hazard Loss Reduction", RAND Report, 2003, 65pp.

APPENDIX C: CURRENT AGENCY ACTIVITIES

	NIST	*NSF*	*NOAA*	*FEMA (DHS)*	*NASA*	*USACE*	*FHWA*
Understanding, Predicting and Forecasting							
Enhancing knowledge, information and data on severe winds	Uniform description of wind increase with height; work on wind turbulence; post-event documentation and damage analyses.	Better understanding of atmospheric dynamics of straight- line winds	Wind observing systems and analysis	Post-event building performance assessments	Experimental and modeling research to improve understanding of severe storm development and lifecycle	Effects of wind on water and storm surge	Monitored and evaluated wind conditions at bridge sites; deployed weather stations along highways
Improving prediction of hazardous wind events	Developed extreme wind climatological models	Tornado genesis and tornadic vortex structure	Wind speed and direction forecasts		Work with NOAA to develop new technologies and use existing scientific satellite observations to improve wind storm prediction		Evaluated wind data and characterized wind and storm conditions at bridge sites
Understanding and quantifying wind loading	Developed novel static and dynamic wind loading estimation techniques	Improved understanding of fundamen-tal physics that control hurricane intensity		Post-event building perfor-mance assessments			Measured wind conditions and structural response; studied wind/structure interaction
Understanding the perception of wind hazard risk		Studies of hazard risk perception		Outreach and training	Satellite observations provide global context to illustrate wind storm hazard risk to the public		Studied user/owner perception of risk
Mapping wind hazards	Developed techniques for improved extreme wind speed maps	Investigation of straight-line winds	Wind speed and direction analysis for input to HAZUS and other tools		Satellite observations from Landsat, EOS provide capability to map potential high risk regions		

Appendix C. (Continued)

	NIST	*NSF*	*NOAA*	*FEMA (DHS)*	*NASA*	*USACE*	*FHWA*
Assessing Impacts Investigating wind resistance of buildings, structures and critical infrastructure	Pioneered studies of structural ultimate capacity and dynamic response under fluctuating wind loads. Post-event damage analyses.	Rapid-response field investiga-tions of structural and societal damage		Post-event building perfor-mance assessments			Conducted aerody-namic assessment of new bridge designs in the laboratory; evaluated structural perfor-mance through full scale measurements
Developing improved tools for component- and structure-level simulation and numerical modeling of wind effects	Developed technology for computer-intensive use of pressure time histories for structural design	Development of instrumentati on for the observational study of atmo-spheric winds					Developed methods for simulations in laboratory; developed numerical models for predicting bridge response
Developing improved tools for loss assessment of wind hazards	Assisted the State of Florida in developing such tools	Design analyses for storm surge, wave actions, and wind speed on structures		Developing hurricane loss estimation model (HAZUS)	Ground, air and satellite based scientific observations provide capability to assess pre- and post-storm impact	Empirical models for debris volume estimation	
Assessing social costs		Economic analyses of wind disaster impacts		Benefits of mitigation study; developiong indirect economic loss model			
Reducing Impacts							
Assessing and communicating risk	Incorporated effect of uncertainties on safety margins for wind loads	Basic research on warning and risk communicati on	Airborne photos and Lidar data for damage assessment; watches and warnings for high wind events; NOAA all hazards radio	Developing hurricane loss estimation model (HAZUS); Design and construction guidance			Developed new procedures for performing aerodynamic assessments

	NIST	*NSF*	*NOAA*	*FEMA (DHS)*	*NASA*	*USACE*	*FHWA*
Developing prototype structural requirements		Improved modeling of wind effects upon structures		Consensus design standard input		Guidelines and requirements for standing seam metal roofing, building survivability, and optimized design	Identified details important to structural performance in winds
Demonstration, education, training and outreach on improved codes and building guidelines	Participates in the ASCE 7 Standard Committee on Wind Loads; trains future wind engineers	Research on Code enforcement and code design analyses		Develop design and construction guidance; Outreach and training		Snow fence construction to reduce wind impacts in cold regions	Demonstrated issues to visitors at laboratory; presented findings at major conferences
Guidance on retrofitting	Provides research; sponsored workshop	Analyses of effectiveness of structural retrofitting		Develop design and construction retrofit guidance			Evaluated structures demonstrating wind problems; developed retrofit solutions
Innovative technologies	Developed Database-Assisted Design methodologies; estimated ultimate capacities through non-linear analysis	Structural health monitoring				Wind noise mitigation	Investigated streamlining of structural form, aerodynamic treatment of structural surfaces, use of dampers and damping materials, use of active/passive control systems
Land use measures and cost effective construction practices		Research on effectiveness of non-structural mitigation		Develop/promote wind-resistant construction guidance		Design guidance on structural issues	
Preparedness and Enhancing Community Resilience							
Developing tools for community preparedness to wind hazards		Research on community emergency preparedness and response capabilities,	Storm Ready community program	Develops tools for effective State and local evacuation and sheltering plans; implementation and training for			

Appendix C. (Continued)

	NIST	*NSF*	*NOAA*	*FEMA (DHS)*	*NASA*	*USACE*	*FHWA*
				design and construction guidance including HAZUS			
K-12 and college education needs	Trains future wind engineers	Research on emergency management education	Tornado and hurricane educational materials	Develop public outreach materials; materials for university courses; HAZUS training	NASA Earth science educational outreach programs increase understanding and awareness of wind storm hazards and impact		Conducted demonstrations and tours for students visiting the laboratory; presented seminars at universities as requested
General public awareness and outreach	Issues recommend-dations for wind engineering research	Surveys of public risk perception and knowledge of protective actions	Pre-tornado and pre-hurricane season public awareness campaigns	Develop public outreach materials; materials for university courses; HAZUS training			
Evacuation Planning		Evacuation Research and Decision-Tool development		Develops tools for effective State and local evacuation and sheltering plans		Cooperative evacuation route development	Participate with the State Highway Agencies in evacuation planning for all hazards
Enhancing disaster-resistance of Building Codes and Standards		Code analysis and studies of enforcement		Facilitate improvements to national model building codes and standards			
Building public and private partnerships				Maintaining and enhancing partnerships with Agencies, Universities, and private sector		Involvement with TISP and other public-private groups	
Conducting emergency response exercises		Basic research on emergency response		Emergency response and preparedness exercises		Active exercise program	Conduct regular emergency response

	NIST	*NSF*	*NOAA*	*FEMA (DHS)*	*NASA*	*USACE*	*FHWA*
				thru FEMA Prepared-ness Division			exercises for all hazards

APPENDIX D: AGENCY SUMMARIES

National Institute of Standards and Technology

Understanding, Predicting and Forecasting

NIST researchers have proposed a uniform description of the variation of wind speeds with height to supersede two descriptions currently in use: the logarithmic law, used by meteorologists, and the power law, used by the ASCE 7 Standard. The current dichotomy makes it difficult to transfer knowledge from the meteorological community to the structural engineering community. NIST has also performed work on wind turbulence spectra, and has shown that efforts to develop overly detailed turbulence descriptions may be unnecessary for a broad class of structural calculations.

NIST conducts research aimed at developing extreme wind climatological models, both non-directional and directional, aimed at estimating extreme speeds with long mean recurrence intervals.

NIST develops and uses innovative computer-intensive, user-friendly methods to quantify wind loading, e.g.: NIST has developed database-assisted design, an advanced approach to defining wind loads on buildings which is based on information obtained in wind tunnel tests. This is expected to reduce errors in the estimation of wind loads by up to 50% and to result in stronger structures at lower costs. In addition, these methods are used in conjunction with nonlinear structural analysis techniques to develop the technology required for ultimate capacity-based design that could save up to 10% or even more in construction steel consumption nationwide while reducing expected losses due to winds. This development is of potential interest to those entities with an interest in "greener" building technology and embodied energy reduction. NIST develops techniques for the estimation of wind effects that account realistically for wind directionality characteristics, and develops estimation methods to help assure higher safety levels for tall buildings that experience dynamic effects.

Through its extreme wind climatological work NIST has contributed to efforts to improve the mapping of hazards due to non-hurricane winds, hurricanes, and tornadoes.

Assessing Impacts

NIST investigates wind-induced damage that can provide new knowledge or recommendations for improvements to codes, standards, and practices that will reduce public risk and economic losses in future. The post-disaster investigations conducted by NIST have led to changes in practices, standards, and codes to enhance the health and safety of the American public. Examples include:

- Improvements to the wind speed requirements for coastal and near coastal regions and anchoring provisions for manufactured (mobile) homes in the Department of Housing and Urban Development national regulatory standards following Hurricane Andrew which struck Florida in 1992 and Hurricane Camille which struck Mississippi in 1969.
- An investigation following the Jarrell, TX, tornado of 1997 led to the development of an enhanced Fujita (EF) scale for tornado intensity (used to estimate wind speed based on observed damage to physical structures) by the Working Group for Natural Disaster Reduction/Post-Storm Data Acquisition in the Office of the Federal Coordinator for Meteorology (which includes NIST as a member) and subsequently has received approval by the National Weather Service for implementation during the Spring 2006 severe weather season.

Further, NIST's program for Hurricane Reconstruction in the Dominican Republic after Hurricane Mitch and Georges (1998), in partnership with USAID and HUD, led to the development of (1) a guide for constructing disaster-resistant housing in the informal sector, and (2) a manual for evaluating the disaster resistance of critical facilities (especially concrete buildings). To disseminate this work among practitioners, training workshops were conducted for local builders on the application of the housing guidance and for local engineers on the application of the facility assessment manual.

As part of its broader mission to meet the measurement and standards needs of the building and fire community, NIST is engaged, singly or in cooperation with other institutions, including NOAA, in wind speed measurements, specialized use of existing data, modeling, and prediction. These efforts are aimed at characterizing the wind environment and the wind climatology near the Earth's surface that are needed for realistic engineering assessments of structural safety, dispersion of fire products, and fire growth and spread. NIST also develops software for the user-friendly utilization of NOAA's ASOS data to create sets of wind speeds in formats suitable for extreme value analyses, and develops and disseminates the most advanced techniques for such analyses. The techniques are validated by using Monte Carlo simulations to estimate the effect of non-asymptotic behavior. NIST also assists the ASCE 7 Standard Task Committee on Wind Loads in developing realistic, updated U.S. wind hazard maps.

NIST is part of the team that develops the Florida Public Loss Prediction model at the request of the Florida Department of Financial Services. Through nonlinear analysis, NIST also develops novel algorithms for prediction of incipient wind-induced structural collapse or damage to building envelope components.

NIST has conducted studies of ultimate capacities of structures under fluctuating wind loads which, if continued, will facilitate safer structures while at the same time reducing material costs and embodied energy.

NIST investigates the systematic use of database-assisted design wherein pressure time histories throughout the building envelope are recorded in electronic form for use in the design process. NIST also has the capacity to develop computational fluid dynamics tools for use in wind engineering.

Reducing Impacts

NIST has capabilities in the area of risk assessment, and has worked on the development of appropriate safety margins for wind loads.

NIST is represented and active in the ASCE 7 Standard Committee on Wind Loads, and provides training to several high-school, undergraduate, graduate students, and postdoctoral research associates per year in the area of wind engineering.

NIST has supported the development of retrofitting guidance both through its research and by holding a workshop on the subject.

NIST has developed several innovative technologies in wind engineering, e.g., Database-Assisted Design methodologies, estimation of ultimate capacities through non-linear analysis, methods for simultaneously taking into account the directionality of the extreme wind climate, of the building's aerodynamic and dynamic response.

Preparedness and Enhancing Community Resilience

Through its investigations NIST has issued recommendations for research and implementation of existing knowledge in a variety of wind engineering areas. For example, NIST's investigation of the World Trade Center towers' collapse of 9/11 found that there were discrepancies on the order of 40% between wind tunnel testing laboratories which estimate wind loads on tall buildings. The labs that were compared were independent labs that are widely used by industry. NIST has recommended that nationally accepted performance standards be developed for: (1) conducting wind tunnel testing of prototype structures based on sound technical methods that result in repeatable and reproducible results among testing laboratories; and (2) estimating wind loads and their effects on tall buildings for use in design, based on wind tunnel testing data and directional wind speed data.

NIST has also recommended that an appropriate criterion be developed and implemented to enhance the performance of tall buildings by limiting how much they sway under lateral load design conditions (e.g., winds and earthquakes).

NIST provides training to several high-school, undergraduate, graduate students, and postdoctoral research associates per year in the area of wind engineering.

National Science Foundation

Understanding, Predicting and Forecasting

The National Science Foundation (NSF) invests a considerable amount of resources in support of competitive peer-reviewed research aimed at defining the wind environment physically and statistically. NSF has awarded grants for research related to hurricanes, tornados, cyclones, and wind. This research is mainly undertaken by the Directorates of Engineering, Geosciences and Social, Behavioral and Economic Sciences.

The Division of Atmospheric Sciences at NSF supports research that covers a broad range of activities relevant to wind hazards. Current research activities include: developing improved understanding of the fundamental physics that control hurricane intensity; research on tornado genesis and tornadic vortex structure; investigations of strong straight line winds from thunderstorms. Better understanding of the atmospheric dynamics of straight-line winds was one of the foci of the Bow Echo and Mesoscale Convective Vortices Experiment (BAMEX) held in the Spring of 2003.

Assessing Impacts

In addition to research in atmospheric dynamics, the Atmospheric Sciences Division supports development of instrumentation (e.g. mobile and airborne Doppler radars) that is required for the observational study of atmospheric winds.

Structural Engineering research funded by NSF has examined the immediate impact of hurricane and tornado activity by funding damage assessment teams to visit disaster areas for post-event triage of structures. These studies are valuable in identifying not only specific anecdotal damage examples but also issues that require further study. For example, uneven damage patterns were observed by NSF-funded teams responding to Hurricane Isabel in 2003. The damage patterns suggest that the peak winds speeds were highly variable as the storm moved inward from the Atlantic Ocean. NSF-funded research is currently developing models for the storm surge and wave actions generated by severe storms and hurricanes. Other research is focused on detailed wind tunnel studies of structures under a variety of wind environments to determine the adequacy of current design and modeling approaches.

Social science research at NSF has supported quick-response research efforts to assess community damages, infrastructure disruption, and individual, organizational, and community social, political and economic impacts. The research attempts to develop damage assessment tools for use by emergency response officials.

Reducing Impacts

Development of improved modeling of wind effects on structures is a primary outcome of NSF-funded Structural Engineering research. Present design standards, such as ASCE 7, have benefited greatly by Federally-funded research into wind and its effects on structures. The current versions of these consensus standards are more comprehensive than previous versions. However, at the present time, most large structures including longer span bridges are subjected to detailed wind tunnel studies to determine the impact of winds on the structure and how to optimize their performance. It is the goal of NSF-funded research to improve the modeling and analysis capabilities in the Engineering community so that in the future, these detailed experimental studies will not be required. This will require on-going improvements in computational systems as well as in the modeling and analysis of the aeroelastic effects of wind. Training of the next generation of Wind Engineers is another aspect of NSF funding that is important. As the founding generation of Wind Engineers has retired, it has become an urgent need to continue the training of young Engineers so that the Wind Community will be able to meet the challenged posed by society in reducing wind effects on the built environment. NSF support for universities engaged in Wind Engineering research is a key component in building of the next generation of Engineers.

Preparedness and Enhancing Community Resilience

NSF has a long history of supporting social science research on risk communication, warning systems and processes, evacuation planning and effectiveness, and individual, organizational and community protective actions undertaken prior to impact. In addition, research on factors facilitating societal adoption of structural and non-structural mitigation measures has been supported for over two decades. Research has also been supported that focuses upon such elements of community preparedness and capabilities improving our understanding and developing tools to measure social vulnerability and damage assessment, and household, business and community recovery from wind disasters.

National Oceanic and Atmospheric Administration

Understanding, Predicting and Forecasting

NOAA maintains an investment, on the order of $600M in its operational numerical weather prediction and forecasting activities within the National Weather Service (NWS). These products include wind forecasts and severe weather outlooks up to 2 and 3 days. In addition, NOAA funds research to improve numerical modeling and forecast decision support systems (about $20M). Much of the research is done within NOAA research laboratories in close collaboration with the NWS, and includes work to improve detection, diagnosis, real-time and retrospective analysis, and forecasts of hurricanes, tornadoes and thunderstorms. The NOAA H*WIND product provides a contoured map of analyzed surface winds for hurricanes before, during and after landfall. In support of its modeling, prediction, and forecasting efforts, NOAA supports and maintains an array of surface wind sensors, radar and satellite observation systems, and mobile tornado tracking, observation, and measurement systems, including the WSR-88D Doppler radars operated with the FAA, and the COOP network. In addition, NOAA research develops and tests new observing capabilities, including airborne microwave wind sensors, GPS sounding technology, Doppler wind Lidar, and radar wind profilers. A new extreme turbulence probe uses pressure ports to measure high velocity winds at high frequency, which has been successfully used during 2004.

Assessing Impacts

NOAA provides photographic and on-the-ground damage surveys immediately following a disaster to determine the severity of the impact of each storm and to provide information that assists in response and recovery. These surveys are particularly useful to correlate the measured winds with the degree of damage. The NOAA Sea Grant program funds real-time hurricane wind measurements and instrumented test homes in coastal environments, assessments of building practices and plans in coastal regions that include wind effects, and education and outreach concerning preparedness and mitigation practices.

Preparedness and Enhancing Community Resilience

Within NOAA, the National Weather Service provides numerous informational and educational materials on protection of the general public and property in high wind events, including hurricanes, tornadoes and straight line winds from thunderstorms. This information is repeated in brief during severe weather warnings. The Sea Grant program, through extension agents, offers education and outreach concerning best construction practices and sighting of coastal development.

Federal Emergency Management Agency*

Understanding, Predicting and Forecasting

Mitigation Assessment Team (MAT) studies of building performance following hurricanes have been frequently and successfully used to assess and understand the impacts of wind storms on the built environment. Those lessons learned have been translated into improvements in design and construction practices and changes to the nation's consensus design standards and national model building codes which directly improves the wind-resistance of construction in high-wind areas. Through the use of a state-of-the-art wind field computer model, the HAZUS loss estimation tool allows states and communities to conduct vulnerability analysis which leads to an improved understanding of how windstorms affect the built environment and their risk from those hazards. Also, ongoing outreach and training on wind hazards enhances the knowledge and capability of all groups to make informed decisions.

Assessing Impacts

Detailed results of HAZUS analysis provide communities, emergency managers, planners, and decision-makers with critical information regarding community vulnerability and likely impacts impending or future hurricanes. This information can have a significant impact on the decision-makers who are responsible for determining what, if any; will be taken before, during, and after hurricanes. Specific assessment activities include:

> Conduct post-hurricane building performance assessments Develop HAZUS hurricane wind loss estimation modeling tool Conduct a study on the benefits of mitigation; and Develop an indirect economic loss model for hurricanes.

Reducing Impacts

A critical, and effective, way to reduce the impacts of hurricanes on the built environment is to have strong, disaster-resistant building codes and in place which effectively enforced. FEMA has worked for over a decade to develop partnerships and successfully advocate for the adoption of wind-resistant provisions to the nation's model building codes and standards. A part of that effort includes participation consensus committees such as ASCE 7 (Design Loads for Buildings and other Structures) Wind Load Task Committee and developing design and construction guidance (new and retrofit) for design professionals, State and local officials, contractors, and the public. We also conduct outreach and training to improve understanding and promote the use of wind-resistant design and construction in high-wind coastal areas.

Preparedness and Enhancing Community Resilience

FEMA works to promote the concept of having emergency plans in place concerning what to do and where to go if a warning is received or a hazard is observed. In partnership with NOAA and USACE, FEMA develops the tools and products to develop effective State and local plans for evacuation and sheltering. In addition, we: provide support for the development and exercise of community and regional evacuation planning; develop and implement state-of-the-art design and construction guidance; train and educate using materials designed for communities, designers, and contractors; develop public outreach materials; establish and maintain partnerships, grants and cooperative agreements with a wide range of Agencies, trade organizations, universities, and private-sector groups; and carry out a robust program of emergency response and preparedness exercises through the FEMA Preparedness Division.

Federal Highway Administration

Understanding, Predicting and Forecasting

A significant component of the FHWA wind research program has been the development of technology and capabilities to collect site-specific information on winds and windstorms to gain better understanding of the wind environment. A select number of sites have been instrumented and monitored by the Aerodynamics Laboratory for periods ranging from weeks to decades to characterize the wind conditions and study impact on structures. Roadway Weather Stations have been deployed on many of their major highway routes to monitor wind as well as general roadway conditions.

Assessing Impacts

FHWA is specialized in the assessment of impact of winds and windstorms on the performance and safety of highway bridges, especially long span bridges, and other highway structures. The agency's Aerodynamics Laboratory has been the primary facility in the U.S. for conducting aerodynamic assessment of new bridge designs and evaluations of existing structures sensitive to wind. Many bridge assessments are also performed at private and public laboratories in Canada, and the results are usually reviewed by FHWA staff. The FHWA Laboratory has been active in developing new or enhanced test procedures, tools for predicting structural response to wind, and models for computer simulation of wind/structure interaction. The Aerodynamics Laboratory also performs full scale measurements on highway structures, in conjunction with wind measurements, to gain better understanding of the impact of winds and windstorms on their performance.

Reducing Impacts

The FHWA wind research program has been working toward reducing the impact of winds and windstorms on highway structures. New designs have been evaluated in the wind tunnel to ensure safety and performance, while retrofits have been identified and implemented for structures with wind problems. Some studies have been initiated to conduct benchmark tests on new structures to establish initial performance characteristics. In addition, efforts have been made to promote structural health monitoring and to incorporate monitoring instrumentation into major new structures. Guidelines are being developed to mitigate wind-induced vibration of cables on long span bridges. Through Transportation Research Board's (TRB) National Cooperative Highway Research Program (NCHRP), AASHTO guidelines for wind related design of highway support structures (signs, signals, luminaries, etc.) have been periodically updated. No national guideline or specification currently exists for the wind resistant design of long span bridges. The FHWA wind research program has just initiated a small project to prepare a synthesis of wind load criteria for long span bridges. Studies are planned to address areas of risk and reliability due to all hazards, including wind.

Preparedness and Enhancing Community Resilience

The FHWA wind research program has attempted to provide some training and awareness to students by providing access to and tours of its research facility in McLean, Virginia. A new demonstration wind tunnel has recently been added to the laboratory to assist in this outreach activity. Some seminars have been presented at selected universities as opportunities arise. More than two dozen graduate students have had the opportunity to perform research in our Aerodynamics Laboratory under the umbrella of the Eisenhower Fellowship Program with the hope of stimulating interest in pursuit of a career in Wind Engineering. FHWA has provided demonstrations and tours of its facility to groups of engineers, industry representatives, public officials, and the general public when requested. Findings of our investigations are presented in a variety of venues including workshops, seminars, national meetings, and international conferences. FHWA also engages in developing predictive traffic analysis tools for planning and traffic management to improve mobility in times of crisis. Models are being tested which allows engineers to perform critical transportation infrastructure analyses and evaluate different incident management strategies during and after any event. Additionally it participates with State highway agencies in evacuation planning and response for all hazards.

National Aeronautics and Space Administration

Understanding, Predicting and Forecasting

NASA operates a number of scientific satellites, including the three flagship satellites Terra, Aqua, and Aura, as part of the Earth Observing System. The EOS instruments provide unique global observations of atmospheric, land and ocean processes that can augment the operational observations of NOAA to improve our understanding of the generation and lifecycle of the large storm systems that produce hazardous wind conditions. NASA satellite observations, when assimilated into operational models, can significantly improve the forecast and prediction of these storms. For measurements of high impact and importance NASA works closely with NOAA and DoD to transition these new measurement capabilities into the next generation of operational weather satellites. Beyond EOS, NASA is also developing new technologies and instruments, including space based precipitation radar and Doppler wind Lidar which will further advance our understanding and capability to predict these events. In addition, NASA is developing advanced global models that can be used for weather prediction and climate simulation; operates, plans, and considers the development of satellite observing systems contributing significantly to the understanding and prediction of windstorms, including space-based Lidar wind profile measurement systems, and the Global Precipitation Mission. NASA performs Observing System Simulation Experiments, typically in collaboration with NOAA, to evaluate the potential of proposed observing systems to contribute to improved forecasts and to design the most effective global observing system.

Assessing Impacts

NASA land and ocean observing systems such as Landsat, MODIS, and SeaWIFs provide the 'big picture' space based view of the impact of large windstorms in affected areas. This big picture view can allow rapid assessment of the effects on the built environment and have the potential to be used for planning and deployment of available resources to the most heavily impacted regions. Other unique ground and airborne instruments can be employed to determine the patterns and magnitudes of coastal change caused by erosion and destruction of buildings and infrastructure following a windstorm. For example, through a cooperative research program NASA, NOAA, the U.S. Geological Survey and the U.S. Army Corps of Engineers are exploring the use of innovative airborne laser mapping systems to quantify coastal change along the entire coastline affected by Hurricane Katrina.

United States Army Corps of Engineers

Understanding, Predicting and Forecasting

Following Hurricane Katrina, ERDC was contacted to provide calculations relative to the effects of wind on water and surge, key information that can be leveraged into future wind mitigation studies.

Reducing Impacts

USACE is presently concerned with building survivability, optimized building and facility design, building in hostile (particularly cold) environments, and in hurricanes and storms. Research proposals addressing structural issues related to wind have been developed for all three services.

USACE has developed guidelines and requirements for standing metal seam roofing and the Corps' Engineer Research and Development Center has recently conducted research in areas of wind/noise mitigation techniques and techniques for constructing snow fences to reduce the impacts of wind in cold regions.

Assessing Impacts

Following Hurricane Andrew in 1992, the USACE participated in a reconnaissance effort related to damage at Homestead Air Force Base and lessons learned as a result of that storm. The Corps' Engineer Research and Development Center (ERDC) participated in an effort for federally owned and leased buildings as part of the Seismic Hazard Reduction Program that involved estimating the number of buildings at risk and probable annual loss and the projected cost to mitigate the existing seismic deficiencies in federal buildings. The knowledge gained in the seismic program can be leveraged for a new wind-focused program. A group within USACE is also working on empirical modeling of debris volume created by wind storms and the associated rain.

Preparedness and Enhancing Community Resilience

USACE participates in cooperative evacuation route development and maintains an active program conducting emergency response exercises.

APPENDIX E: LIST OF ACRONYMS / DEFINITIONS

AASHTO	American Association of State Highway and Transportation Officials
AAWE	American Association of Wind Engineers
ASCE	American Society of Civil Engineers
ASOS	Automated Surface Observing System
ASTM	American Society for Testing and Materials
BAMEX	Bow Echo and Mesoscale Convective Vortices Experiment
COOP	Cooperative Observer Program
DOD	Department of Defense
EOS	Earth Observing System
ERDC	Engineer Research and Development Center
FAA	Federal Aviation Administration
FEMA	Federal Emergency Management Agency
FHWA	Federal Highway Administration
HAZUS-MH	HAZUS Meteorological Hazards
H*WIND	Hurricane Wind
HUD	Department of Housing and Urban Development
IBHS	Institute for Business and Home Safety
ICC	International Code Series
MAT	Mitigation Assessment Team
NASA	National Aeronautics and Space Administration
NCHRP	National Cooperative Highway Research Program
NFPA	National Fire Protection Association
NIST	National Institute of Standards and Technology
NOAA	National Oceanic and Atmospheric Administration
NSF	National Science Foundation
OSTP	Office of Science and Technology Policy
USACE	United States Army Corps of Engineers

End Note

* Specific organizational units may vary because of pending reorganization.

In: National Windstorm Impact Reduction Program
Editor: Ralph Simmons

ISBN: 978-1-63484-291-4

Chapter 3

MEASUREMENT SCIENCE R&D ROADMAP FOR WINDSTORM AND COASTAL INUNDATION IMPACT REDUCTION*

NEHRP Consultants Joint Venture

EXECUTIVE SUMMARY

Windstorms, storm surge, and other coastal inundation events caused 4,045 fatalities and property losses of $250 billion (in 2012 dollars) in the United States during the period of 1996-2012. More people are moving toward the nation's coastlines and thus more of the built environment is at risk from hurricanes and nor'easters and/or coastal inundation due to storm surge and tsunamis. In addition, a large part of the central United States is at risk from tornadoes, and all areas of the country can experience damage from other types of windstorms, such as thunderstorms, derechos, and straight line winds. Sometimes effects to a community are so great from the windstorm and coastal inundation events, the community never recovers, as the tax base moves away and businesses and schools stay permanently closed. The current state of practice for engineering design of the built environment to withstand extreme wind and coastal flood events must be improved in order to improve life safety, reduce property damage, and improve the resiliency and sustainability of communities.

This report describes results from a measurement science research and development (R&D) roadmap development process to identify and describe a much-needed problem-focused program to reduce the impacts of windstorms and coastal inundation hazards. This report is also intended to aid and provide focus for the broader wind, coastal, and structural engineering R&D community. The effort was conducted to assist the National Institute of Standards and Technology (NIST) in planning its research and development program related to these hazards over an approximately eight-year time frame. A Project Technical

* This is an edited, reformatted and augmented version of a contracted report, NIST GCR 14-973-13, issued by the U.S. Department of Commerce, National Institute of Standards and Technology, January 2014. The NEHRP Consultants Joint Venture is a partnership of the Applied Technology Council and the Consortium of Universities for Research in Earthquake Engineering.

Committee comprised of practitioners and academics provided guidance for this project. As a starting point, research recommendations from past reports by the National Research Council, National Science Board, National Science and Technology Council, NIST, and other organizations were identified and reviewed. A two-day workshop of industry's leading practitioners, academics, and representatives from key government agencies was convened to identify the highest priority needs that form the basis of this roadmap. The workshop was co-funded by the National Science Foundation (NSF), with the aim to leverage this effort to also identify fundamental research challenges on a longer time frame than NIST usually operates within. A second half-day workshop was held to help refine and prioritize the specific topics identified in the first workshop. The 30 most relevant and needed research and development topics identified through these workshops are summarized in Table 1. The priorities are illustrated by color (legend is at the bottom of the table). The topics are grouped into four categories: (1) hazard identification; (2) hazard-specific loads and effects; (3) building-specific resistance; and (4) hazard-specific performance-based design.

Table 1. Recommended Windstorm and Coastal Inundation Hazard Reduction R&D Topics

No.	Wind Topics	No.	Coastal Inundation Topics
Hazard			
1	Hazard maps for high intensity non-synoptic events, including tornadoes	20	Quantification of flood-borne debris hazards
3	Quantification of wind-borne debris hazards	21	Effects of over-land flow on waves and other coastal inundation hazards
5	Joint description for hurricane wind, storm surge, and wave hazards	22	Characterization of the coastal inundation hazard for low probability events
2	Effects of climate change on extreme wind event characteristics	19	Effects of sea level rise and future shoreline changes on coastal inundation hazards
4	Hazard modification by terrain, topography, and nearby buildings		
Loads and Effects			
6	Pressure coefficients for wind load determination	23	Field data collection of flood loads
7	Computational fluid dynamics (CFD) tools for practitioners	24	Load combinations for simultaneous flood hazards.
8	Rational approach for wind load analysis	25	Validation of flood forces on buildings
9	Modeling and simulation for determination of wind loads due to non-synoptic winds, including tornadoes		
Resistance			
14	Guidelines for wind vulnerability assessment of existing buildings	26	Flood damage functions and fragility curves.

No.	Wind Topics	No.	Coastal Inundation Topics
11	Field measurement techniques and test methods to evaluate condition of existing construction	27	Effects of erosion and scour on deep foundations
13	Methods or models to predict loss of strength due to aging and deterioration	28	Test methods for materials and systems subject to flooding
10	Test methods for resistance to cyclic loading		
12	Explicit factors of safety for materials and assemblies		
Performance-Based Design			
15	Performance levels and acceptance criteria for wind hazards	29	Performance levels and acceptance criteria for buildings subject to coastal inundation
18	Measurement of windstorm resilience and benefits of performance-based design	30	Sound engineering basis for flood resistant design that is not based on flood insurance requirements
16	Analysis procedures for nonlinear system behavior used in performance-based design		
17	Cyber-based tools to support performance-based design		

Legend: Highest priority R&D topic; High priority R&D topic; Medium priority R&D topic

The report also provides details on each recommended research and development topic, including (1) a description; (2) its estimated cost; (3) the estimated time to carry out needed R&D pertaining to the topic; (4) challenges in conducting the needed R&D; and (5) potential R&D solutions. For each topic, example stakeholders, and their respective roles, are also identified, as well as the significant impacts of the R&D to both engineering practice and community resilience, once the R&D is successfully completed and implemented. Also included in the report are explanations of how priorities for the R&D topics were assigned by workshop participants.

The proposed program description includes discussions on budgetary requirements, schedule and benefits. In cost terms, the program starts at a level of $4.0 million per year and quickly grows to $6.0 million per year over eight years. For the nation as a whole, implementation of the proposed R&D program will yield the following major benefits:

1. Reduced loss of life, injury, damage and economic impact following wind storm and coastal inundation events.
2. More resilient communities through faster recovery following wind storm and coastal inundation events.
3. Reduced costs through risk-consistent design and construction of the built environment subject to wind storm and coastal inundation events.

SECTION 1. INTRODUCTION

The National Institute of Standards and Technology (NIST) has a long history of research and development (R&D) in wind engineering and windstorm impact reduction, and has more recently expanded its R&D efforts to include coastal inundation, following Hurricane Katrina in 2005. With growth in these activities and with increasing national understanding of the significance of windstorm and coastal inundation disasters on the national ability to respond to and recover from these events, NIST has determined there is a need for a detailed *measurement science*[1] roadmap that identifies critical windstorm and coastal inundation impact reduction research and development priorities, to assist NIST with planning its research over the next decade. This roadmap will also aid and provide focus for the broader wind, coastal, and structural engineering R&D community.

1.1. Background

From 1996 to 2012, windstorms and coastal inundation events caused 4,045 fatalities in the United States and U. S. property losses of $250 billion (in 2012 dollars)[2]. More people are moving toward the nation's coastlines and now 40% of the U.S. population lives along the coasts (NOAA, 2013b). Thus, much of the built environment in coastal areas is at risk from hurricanes and nor'easters and/or coastal inundation due to storm surge and tsunamis. In addition, a large part of the central United States is at risk from tornadoes where wind speeds can significantly exceed those from hurricanes, and all areas of the country can experience damage from other types of windstorms, such as thunderstorms, derechos, and straight line winds. Impacts from these windstorm and coastal inundation events are not limited to fatalities, injuries, and property damage, but also include closed schools and businesses, people out of work, and damaged infrastructure, all making community recovery a long and expensive process. Sometimes, effects to a community are so great, the community never recovers as the tax base moves away and businesses and schools stay permanently closed. The current state of practice for engineering design of the built environment to withstand extreme wind and coastal flood events must be improved in order to improve life safety, reduce property damage, and improve the resiliency and sustainability of communities.

NIST research related to windstorm and coastal inundation impact reduction is part of the Engineering Laboratory's Structural Performance under Multi-Hazards Program, whose objective is "to develop and deploy advances in measurement science to enhance the resilience of buildings and infrastructure to natural and manmade hazards." As one of the agencies of the National Windstorm Impact Reduction Program (established by the National Windstorm Impact Reduction Act of 2004 (Public Law 108-360)), NIST also has specific responsibilities to "support research and development to improve building codes and standards and practices for design and construction of buildings, structures, and lifelines."

1.2. Roadmap Scope and Purpose

The scope of the roadmap addresses impact reduction from (1) damaging windstorms, such as hurricanes, tornadoes, thunderstorms, derechos, and straight line winds, and (2) coastal inundation, caused by storm surge and tsunamis. High priority needs for both industry practice and for standards and model codes are included in the roadmap.

The roadmap is intended to help guide NIST's investment in windstorm and coastal inundation impact reduction research. The rationale for public investment in these areas includes loss reduction potential (reduced fatalities, injuries, property loss, and business interruption; continuity of operations; community resilience) and enabling innovation in materials, design, construction, and retrofitting through performance- based engineering, standards, and model codes.

1.3. Roadmap Development Process

The starting point for the roadmap development process was an initial effort to identify recommendations from the research needs literature. A two-day workshop of invited specialists in windstorm and coastal inundation was then convened in June 2012 to identify the highest priority research and development needs, which were refined at a follow-up half-day workshop in October 2012. This report provides a synthesis of the information gathered through the review of the literature and both workshops. A Project Technical Committee consisting of several leading practitioners and academics in the areas of wind engineering and engineering for coastal hazards provided guidance throughout the roadmap development process.

Reports identified during the review of research needs literature of the past 20 years related to windstorm and coastal inundation impact reduction included publications by the National Research Council, National Science Board, National Science and Technology Council, NIST, and other organizations. A total of 20 publications were reviewed, including the traceable documents listed below.

- *Wind and the Built Environment, U.S. Needs in Wind Engineering and Hazard Mitigation*, National Research Council, 1993;
- *Windstorm Impact Reduction Implementation Plan*, National Science and Technology Council, April 2006;
- *Hurricane Warning: The Critical Need for a National Hurricane Research Initiative*, National Science Board of the National Science Foundation, January 2007;
- *Grand Challenges for Disaster Reduction*, National Science and Technology Council, Subcommittee on Disaster Reduction, 2008;
- *Disaster Resilience: A National Imperative*, National Academies Press, 2012; and
- ATC-57, *The Missing Piece: Improving Seismic Design & Construction Practices*, Applied Technology Council, 2003.

The ATC-57 report, which was formulated by the earthquake engineering community with funding in part from NIST, provides a research and development roadmap that was studied as a possible format for this measurement science roadmap on windstorm and coastal inundation. The list of documents reviewed at the outset of the roadmap development process (see Appendix A) is organized around two subject areas and five program elements that were patterned after those in the ATC-57 report and initially described in the work plan for the development of this roadmap to reduce the impacts of windstorm and coastal inundation hazards.

The literature provides an additional perspective on the research issues that is not possible from workshops or other data collected over a short period of time. The effort uncovered the fact that there have been more research needs assessments related to wind engineering than for coastal inundation. The documents on coastal inundation have mainly been produced in the last several years, in part as a consequence of the devastation wrought by storm surge during Hurricane Katrina in 2005.

The two-day workshop in June 2012, convened to identify the highest priority needs that form the basis of this roadmap, was attended by the industry's leading practitioners, academics, and representatives from key government agencies. The workshop was co-funded by the National Science Foundation (NSF), with the aim to leverage this effort to also identify fundamental research challenges on a longer time frame than that in which NIST usually operates. The workshop was held at the headquarters of the American Society of Civil Engineers (ASCE) in Reston, Virginia. Participants were provided with the information from the review of the research needs literature prior to the workshop.

The design of the June 2012 workshop was such that all 61 participants were able to contribute by addressing the research needs in multiple ways. Initially, the participants were asked what their future vision of a resilient community was and describe how that resilient community could be defined. They were also asked to identify any gaps in current knowledge that need to be addressed to enable disaster- resilient communities. The participants were then grouped by technical disciplines and asked to identify research needs and gaps in the current state of the art and state of the practice for determination of windstorm and coastal inundation hazards, loads, resistances, and performance-based design.

During the June 2012 workshop, participants in each of the five breakout groups (one group per program element) were asked to identify the top ten research needs for each roadmap program element. The participants were then asked to rank the top three most important topics for each element. Results from the June 2012 workshop were reviewed, discussed, and refined at the October 2012 workshop. Participants in the latter event were also asked to identify the research and development topic they considered the most important and the topic they considered the best value. Based on the input received from both workshops, the research and development project topics were then revised or combined to eliminate duplication, overlap, and expand topics that were too narrowly focused, resulting in a finalized set of 30 priority topics.

1.4. Roadmap Framework

The framework adopted for the measurement science roadmap for wind and coastal inundation impact reduction consists of four key elements used to describe the range of research that needs to be pursued for both hazards. The recommended research and development efforts include work in hazard identification, hazard loads and effects, building-specific load resistance improvement, and developing performance-based design methodologies. Each area of research and development is briefly described below.

Hazard Identification

The need to improve methods for identifying and quantifying both the windstorm and coastal inundation hazard was described by workshop participants as one of the most important elements of the research program. It is extremely difficult to define accurately the loads created by the hazards if the hazard is not adequately identified. Improving hazard identification requires research into methods to better determine the extent and magnitude of the hazard, such as the probable inundation levels and wave heights for 500-year and larger coastal inundation events, and to identify elements of the hazard that are not currently described or mapped, such as non-synoptic winds or flood-borne debris. The effect of future conditions such as sea level rise must be included in the identification of the hazard, and it is also important to identify the hazards on the same risk-basis so designs for the hazards are risk-consistent.

Loads and Effects

The workshop participants identified gaps in the knowledge of loads and effects caused by the full spectrum of windstorm and coastal inundation hazards, ranging from the need for improved wind load external pressure shape factors and improved methods for determining wind loads, to needing reliable methods for determining coastal inundation loads caused by moving water and water- borne debris. Improving knowledge of loads and effects also includes the need for developing methods of load determination for non-synoptic winds and for developing computational methods and tools not sufficiently developed for use by the practice.

Improved Building-Specific Load Resistance

The knowledge about the actual resistance of buildings and the components to windstorms and coastal inundation is limited to that provided by testing products, from testing structural assemblies, and from post windstorm and coastal inundation-event analysis. There is a very large gap in knowledge about the resistance of existing buildings to these hazards and little information about the effects of aging on wind and coastal inundation resistance. Satisfactory building design for these hazards requires an adequate load determination methodology and an adequate approach to determine resistance to the loads.

Performance-Based Design

There has been little research and development work in performance-based design for wind loading and almost no performance-based design R&D work for coastal inundation. The suggested approach for this research and development element is to build on existing capabilities in performance-based seismic design, which is well developed, by developing performance targets and reliability criteria for the hazards considered by this roadmap, for both new and existing buildings. Building performance during windstorms and coastal inundation should be linked to community resiliency.

1.5. Roadmap Organization

This roadmap is organized as follows. Section 2 describes the vision for windstorm and coastal inundation resilient communities that is based on input received from the June 2012 workshop participants. Section 3 describes the "Grand Challenges" in windstorm and coastal inundation loss reduction, as identified by the June 2012 workshop participants. Details of the 30 priority research project topics that rose to the top of the many projects discussed and debated at both the June 2012 and October 2012 workshops are provided in Section 4. Section 5 provides a prioritization of the proposed research topics, along with a summary of the costs and duration of the associated projects. Four appendices are also included that provide details about the workshops and the review of the literature. Appendix A identifies research needs and recommendations from the review of the literature and mapped to the five roadmap program elements that had been identified in the original project work plan.

Appendix B contains the June 2012 workshop agenda, participant list, description of technical presentations, and web site address for accessing the presentations from the workshop. A description of the October 2012 workshop and a list of individuals who attended are provided in Appendix C. The priority research and development needs, as identified by the June 2012 workshop participants and mapped to the initial set of five roadmap program elements, are provided in Appendix D. Following the appendices are acronym definitions and references.

SECTION 2. VISION FOR WINDSTORM AND COASTAL INUNDATION RESILIENT COMMUNITIES

The term 'resilience' has a number of definitions. With regards to hazard events, and responding to hazard events, it has been defined as "the ability to prepare and plan for, absorb, recover from, and more successfully adapt to adverse events" (National Academies Press, 2012) or, similarly, "the ability to prepare for and adapt to changing conditions and withstand and recover rapidly from disruptions" (Presidential Policy Directive PPD-21 (Obama, 2013)). Resilience includes the ability to withstand and recover from deliberate attacks, accidents, or naturally occurring threats or incidents.

Community resilience can address a range of topics that include physical security, business continuity, emergency planning, hazard mitigation, and how the built environment (e.g., facilities, transportation, utilities) resists and rapidly recovers from disruptive events. With regards to buildings and infrastructure systems in communities, resilience can be described as physical systems that resist damage through adequate strength and mitigation measures, and that can be readily repaired so that important community functions are maintained during and after a hazard event. To achieve such performance, goals and criteria for resistance and recovery from windstorm and coastal inundation events need to be developed at the facility, system, and community level.

2.1. Resilience

The characteristics of a resilient community include preparation for hazard events, continuity of operations during and after hazard events, and recovery plans that allow the population to resume work and activities within a reasonable period of time.

A resilient community has performance goals for primary hazard events that address community development, mitigation, operations, and recovery. The community performance goals provide a consistent framework for defining specific criteria for facilities and infrastructure systems. Government agencies and critical functions, such as police, fire, emergency response, and healthcare, should be able to provide service immediately after the event with no significant interruption or decrease in service levels. Both public and private sector organizations should have disaster response plans consistent with the community performance goals to increase the likelihood of less down time, continued employment, and lower recovery costs.

The issue of resilience in the built environment was addressed in several NIST-sponsored workshops in 2011 and are summarized in a NIST Technical Note entitled, *Developing Guidelines and Standards for Disaster Resilience of the Built Environment: A Research Needs Assessment* (NIST, 2013). The technical gaps and research needs for developing guidelines and standards supporting community resilience were characterized as short term (less than three years) and long term (greater than three years) and are listed below:

Short term activities:

- Identify technical gaps and research needs from reviews of past disaster events and existing model codes and standards;
- Define resilience terminology for the built environment; and
- Develop guidance for community resilience planning.

Long term activities:

- Develop significant hazard definitions and multiple hazard levels;
- Develop risk-consistent performance goals and metrics;
- Develop tools and metrics to support quantitative technical assessment, policy development, and decision making; and
- Develop codes and standards that address resliency.

The short term activities address resilience planning at the community level, where the role of facilities or systems in a community's physical resilience is defined. The long term activities address identification and development of tools and metrics that determine the expected level of performance during and after a hazard event, and include recovery of functional performance.

2.2. Vision

The vision of a resilient community was discussed by the June 2012 workshop participants. The themes of that vision are:

1. The community understands hazard risks, has an effective risk management program, and plans for resilience across the community through mitigation actions, disaster procedures, and plans for recovery.
2. The community can sustain the effects of a windstorm or coastal inundation event, provide basic human needs immediately after the event, and return to normalcy within a short period of time after the event with little economic and social disruption;
3. The population, tax base, and business activity returns to normalcy within a short period of time after the event, indicating community functionality and stability.

The following gaps in current knowledge, information, and resources were also identified by the workshop participants:

- Degradation in facilities and infrastructure systems due to age and deterioration are not addressed by codes and standards. The range of capacity in physical systems across a community needs to be considered in resilience planning and performance goals.
- Climate change needs to be considered when setting resilience performance goals in areas where it may affect the magnitude and extent of windstorms and coastal inundation.
- Design hazard information with a probabilistic basis for tornadoes, tsunamis, and storm surge events needs to be developed to enable communities to develop risk-consistent buildings and infrastructure systems.
- Cost analysis tools are needed that can address long-term resilience strategies. Codes and standards for buildings primarily focus on life safety issues, and address minimum requirements. Codes and standards for utilities primarily focus on reliability of service. While codes and standards are necessary for a resilient community, they may not be sufficient to reach the desired level of resilience.
- The private sector needs to be engaged in creating a resilient community.
- The public needs to be better informed about the importance of resilience and the steps needed to improve a community's ability to respond to and recover from a windstorm or coastal inundation event.

The ability to communicate risk to the public is critical, yet has proven for the most part to be ineffective. For individuals, businesses and government agencies to work together towards a resilient community, the risks of building and living in coastal environments must be clearly conveyed. Based on these gaps, the following four areas were identified as the most significant gaps in available information and tools for communities that want to develop resilience goals and plans:

1. **Measurement Techniques Needed**. Metrics for assessing community resilience must support decision making about priorities and associated levels of resilience. Resilience metrics need to address individual buildings and neighborhoods, infrastructure facilities and systems, system interdependencies, recovery time. Metric development is expected to be a long-term effort and is key to developing resilient communities.
2. **Risk Communication Required**. Risk communication to the public, private, and business populations is required to shift the disaster recovery paradigm, where government is expected to address planning and recovery, to more proactive, community-wide disaster planning and loss reduction. A key element of risk communication is clear statements about the hazards, associated risks, and community goals and plans.
3. **Community-Scale Solutions Needed**. Community-scale tools and examples that provide a framework for considering the many issues involved in planning resilience, including reducing losses from disasters and speeding up recovery, are needed.
4. **Changing the Development Paradigm**. Given the large uncertainty in predicting windstorm and coastal inundation intensity and damage, coastal communities might consider policies that identify areas that should not be developed or abandon their development, particularly for coastal inundation.

2.3. Connecting Resilience to Measurement Science Roadmap

The projects identified in this roadmap will help fill some of the gaps identified above. The projects that focus on hazard identification are key to developing metrics that could measure community resiliency, and they are vital in determining how to communicate hazard risk within communities. The projects that address long-term climate issues and those that improve building performance over a long time frame are at the heart of resiliency and are crucial to improving a community's chance at more rapid recovery.

The roadmap topics do not address all of the gaps, however. While thorough hazard identification is fundamental to reducing risk uncertainty and important to risk communication, acting on the information, determining how, when and where to communicate risk, and how to mobilize communities to work toward improving their own long-term resiliency is a matter of public policy and is beyond the issues addressed in this roadmap. Any community decision to abandon development, or even severely restrict it, would be a matter of public policy and would require lengthy public debate, even though the fundamental hazard identification and quantification would hopefully be the basis for those policy discussions.

SECTION 3. GRAND CHALLENGES IN WINDSTORM AND COASTAL INUNDATION IMPACT REDUCTION

The workshop participants identified "Grand Challenges" for both windstorms and coastal inundation research needs. These topics are anticipated to require a longer time frame (on the order of a decade or more to fully address) than the issues articulated in the remainder of this roadmap, but help shape a comprehensive research agenda for which new or additional knowledge is sought. However, the roadmap includes many topics on which significant progress could be made towards meeting these challenges.

Workshop participants also identified one means for efficiently making progress on research objectives for both windstorm and coastal inundation issues—development of a laboratory network for wind and coastal inundation-focused research similar to the Network for Earthquake Engineering Simulation (NEES)[3]. Such a network could provide shared-use laboratory facilities and allow research institutions to improve collaboration, with an objective to maximize effectiveness and efficiency of research across both the windstorm and coastal inundation hazards.

The Grand Challenges identified are focused in four main areas of study. They are:

- Developing new computer-aided tools and techniques to better simulate and predict hazard effects
- Developing techniques to measure the response of buildings to hazards and using the information as predictive tools
- Incorporating future conditions into hazard identification
- Improving community-wide resiliency and resistance to the effects of the hazards

Following are the identified Grand Challenges for both windstorm and coastal inundation hazard reduction. The Grand Challenges are not listed in priority order.

3.1. Windstorm Grand Challenges

3.1.1. Develop Simulation Techniques for Non-synoptic Windstorms

Physics-based simulation methods that accurately represent the near-surface wind field during non-synoptic windstorms (e.g., thunderstorms, tornadoes) are needed. Applications of these simulation methods include improved hazard quantification and mapping, enhanced computational fluid dynamics modeling of wind loads and wind structure interaction, and more accurate estimation of windstorm disaster losses.

3.1.2. Define Hazard Characteristics of Wind Interacting with Other Environmental Effects

Windstorms most often are accompanied by other environmental effects, such as storm surge, rain, and wind-borne debris. The influence on building designs and event losses caused by the simultaneous impact of wind and one or more of these other environmental effects is not well understood and has been inadequately considered in current design practices. Studies have been initiated on the effect of high winds and wind-driven rain, yet much is to be learned about this means of water intrusion, as the combined effects of high winds with storm surge at a building site have not been studied in any substantive way. Results from research studies of the combined effects of these hazards could significantly improve designs, suggest changes in materials being specified, and reduce losses.

3.1.3. Develop Computer-Based Simulation Tools for Use in Design

Computer simulation tools such as computational fluid dynamics are needed to better understand wind pressures on buildings and other structures. Computational fluid dynamics, coupled with models of the structural response of the building, can be used to predict building performance from a high-wind event. Development of these simulation capabilities could ultimately enable a wind tunnel simulated on a computer desktop with computational fluid dynamics as a core function. Integrating such a simulation capability with Building Information Modeling (BIM) of a particular building would enable design of that building based on wind loads specific to each building and surrounding terrain. Variations in the extreme wind climate could be simulated, executing computational fluid dynamics simulations of the building and surroundings for multiple storms and directions, and then evaluating loads and load effects for design, including incorporating the effects of aging and deterioration.

3.1.4. Develop 'Smart' Buildings

'Smart' buildings would have wind performance monitoring systems built into the structural system that could provide information on building health or damage, and assist in predicting possible collapse, warning any building occupants of impending dangers, predicting excessive drift caused by wind, or alerting an owner about infiltration of excessive amounts of water or air. These buildings could lower storm shutters or secure photovoltaic solar panels prior to a storm impacting the building.

3.1.5. Improve Infrastructure Performance during High Wind Events

Power transmission and distribution systems are often crippled by high wind events. Methods to transmit power other than by electric lines draped between poles where both wind speeds and fallen trees disable the power grid are needed. Possible options might be to transmit power using electronic transmitters, moving the power lines underground or using the network of roads and streets to transmit the power within inlaid electrical conductors. Improved performance of electric lifelines would have a significant positive impact on reducing losses from windstorms. Similar performance of telephone and IT distribution systems has been observed during high wind events and solutions to these problems could be similar to those pursued for electric distribution systems.

3.1.6. Develop Community-Wide Wind-Resistant Design Techniques

Techniques are needed for determining wind vulnerabilities across a range of contiguous communities to make it more feasible to develop strategies to reduce those vulnerabilities. These techniques would include methods for creating redundancies between community systems such as infrastructure, critical facilities, and housing and employment facilities, so that communities could address their most urgent needs. These techniques would go well beyond the need for adoption and enforcement of stronger building codes.

3.2. Coastal Inundation Grand Challenges

3.2.1. Develop Observation and Measurement Technologies for Coastal Flood Hazard, Load, and Building Response Data Collection

Coastal inundation hazard data include time histories of water elevation, wave height and velocity, amount and type of sediment, and amount and characterization of flood- borne debris. Technologies or measurement systems are needed where buildings are located so information can be captured on water velocity, wave, and sediment supply to inform design approaches. Technologies might include load sensors attached to buildings to collect this information or buildings instrumented when built to continuously provide relevant coastal hazard information. The hazard data collected is extremely important in improving the understanding of overland flow of storm surge, the magnitude of flood loads on buildings, and the response of the buildings to those loads.

3.2.2. Create Modeling Tools to Predict the Effects of Overland Hydrodynamic Transport

During an extreme flooding event, loads on buildings and other structures are affected by a variety of phenomena, such as significant near-shore flood effects, including waves and storm surge; flow through the built and natural environments, considering time dependency and building interaction; sediment transport, including the effects of scour and erosion; and generation and transport of debris. Understanding the overland flow and the propagation of waves inland is important to determining loads on buildings and would help define the flood hazard for communities. This knowledge could be used to create powerful predictive tools to estimate potential losses due to overland hydrodynamic transport.

3.2.3. Develop Coastal Inundation Hazard Models Capable of Resolution at Parcel-Level Scales

A physics-based flood mapping methodology is needed that will provide both inundation limits and elevations at any building site. The methodology should include all coastal influences, as well as rainfall runoff, and be available to users electronically at the parcel level. Hazard information at the parcel-level would provide valuable flood design information for any particular site.

3.2.4. Develop Probabilistic Storm Surge and Tsunami Models that Incorporate Changing and Future Conditions

Methodologies are needed to determine probabilistic flood risks for short- and long-return periods for both hurricanes and tsunamis, considering changing and future conditions. These conditions include sea level rise and shoreline changes, including long term erosion and any other effects of climate change on water surface elevations and land forms. The changing conditions should also include those that could generate a tsunami, such as underwater landslides or shifting tectonic plates under the ocean floor, or changes in bathymetry that would affect tsunami surge.

3.2.5. Develop Innovative Flood Resistance Techniques for Buildings

Current coastal inundation hazard mitigation techniques are limited to elevating structures above some predicted event or relocating them away from such an event. Techniques must be developed that improve building performance from coastal inundation hazards that are more comprehensive than those represented by simply elevating or relocating a structure. Pursuit of this challenge requires extensive knowledge about the magnitude of the coastal inundation hazard elements (e.g., changes in water elevation, wave height and velocity, amount and type of sediment, and amount and characterization of flood-borne debris) and about overland hydrodynamic transport, as well as accurate mapping of the hazard for various return periods, including inundation limits and elevations at specified building sites and considering changing conditions with time. Innovative materials and/or building systems to better withstand flood forces and allow faster recovery would significantly impact a reduction in coastal inundation losses.

3.2.6. Develop Community-Wide Flood-Resistant Design Techniques

Techniques are needed for determining flood risks and vulnerabilities across communities to make it possible to develop strategies to reduce those vulnerabilities, based on the risks. These techniques would include methods for creating redundancies between community systems, such as infrastructure, critical facilities, housing and employment, so that communities could address their most urgent needs. These techniques would go well beyond the need for adoption and enforcement of stronger building codes or floodplain ordinances.

SECTION 4. RECOMMENDED RESEARCH AND DEVELOPMENT TOPICS

4.1. Introduction

The research and development (R&D) project topics recommended by participants in the June 2012 and October 2012 workshops, and the subsequent effort to eliminate overlap and duplication and refine their focus, resulted in the identification of 30 priority research and development topics. The topics were then analyzed, categorized, and further elucidated for inclusion in the overall framework for this measurement science R&D roadmap for windstorm and coastal inundation impact reduction.

The priority R&D topics were grouped into two primary categories (wind hazard reduction and coastal inundation hazard reduction), with the following sub categories: (1) hazard identification; (2) hazard-specific loads and effects; (3) building-specific resistance; and (4) hazard-specific performance-based design (see Table 4-1). Workshop participants also examined possible research topics by the five program areas identified in the original project work plan from several different directions, including looking for grand challenges, exploring resiliency issues, and suggesting projects with high impacts and rapid delivery times across disciplines (see Appendix D).

Table 4.1. Priority Research and Development Topics, as Identified by Workshop Participants

No.	Research and Development Topic
Wind Hazard	
1	Hazard maps for high intensity non-synoptic events, including tornadoes
2	Effects of climate change on extreme wind event characteristics
3	Quantification of wind-borne debris hazards
4	Hazard modification by terrain, topography, and nearby buildings
5	Joint description for hurricane wind, storm surge, and wave hazards
Wind Loads and Effects	
6	Pressure coefficients for wind load determination
7	Computational fluid dynamics tools for practitioners
8	Rational approach for wind load analysis
9	Modeling and simulation for determination of wind loads due to non-synoptic winds, including tornadoes
Wind Resistance	
10	Test methods for resistance to cyclic loading
11	Field measurement techniques and test methods to evaluate condition of existing construction
12	Explicit factors of safety for materials and assemblies
13	Methods or models to predict loss of strength due to aging and deterioration
14	Guidelines to wind vulnerability assessment of existing buildings.
Performance-based Design-Wind	
15	Performance levels and acceptance criteria for wind hazards
16	Analysis procedures for nonlinear system behavior used in performance-based design
17	Cyber-based tools to support performance-based design
18	Measurement of windstorm resilience and benefits of performance-based design
Coastal Inundation Hazard	
19	Effects of sea level rise and future shoreline changes on coastal inundation hazards
20	Quantification of flood-borne debris hazards
21	Effects of over-land flow on waves and other coastal inundation hazards
22	Characterization of the coastal inundation hazard for low probability events
Coastal Inundation Load and Effects	
23	Field data collection of flood loads
24	Load combinations for simultaneous flood hazards
25	Validation of flood forces on buildings
Coastal Inundation Load and Effects	
26	Flood damage functions and fragility curves
27	Effects of erosion and scour on deep foundations
28	Test methods for materials and systems subject to flooding
Performance-based Design-Coastal Inundation	
29	Performance levels and acceptance criteria for buildings subject to coastal inundation
30	Sound engineering basis for flood resistant design that is not based on flood insurance requirements

Analysis and further elucidation of the priority R&D topics identified in Table 4-1 resulted in the expansion of each topic to include (1) a more in-depth description; (2) an indication of the roadmap program element in which it falls; (3) its estimated cost; (4) the estimated time to carry out needed R&D pertaining to the topic; (5) challenges in conducting the needed R&D; and (6) potential R&D solutions. One- page summaries for each priority research topic containing the above-cited information are provided throughout the remainder of this section. For each topic, example stakeholders, and their respective roles, are also identified, as well as the significant impacts of the R&D to both engineering practice and community resilience, once the R&D is successfully completed and implemented.

The costs are based on estimates provided by the workshop participants, based on their knowledge of costs for similar R&D efforts. Estimated costs for each research and development topic are provided using one of the following ranges: $100,000 to $500,000 (a low cost); $500,000 to $1,000,000 (a moderate cost); $1,000,000 to $3,000,000 (a high cost); and exceeding $3,000,000.

Similarly, the time requirements to properly address each R&D topic was estimated by the workshop participants, based on their personal experience with comparable R&D efforts. Estimates are provided using the following time period ranges: 1-to-2 years (a short time period); 2-to-5years (a moderate time period); 5-to-8years (a long time period), and exceeding 8 years.

4.2. Priority Windstorm R&D Topics

The priority windstorm R&D topics are elucidated in the following four subsections: identification and characterization of wind hazards (Section 4.2.1); wind loads and effects (Section 4.2.2); wind resistance (Section 4.2.3); and performance-based engineering methodology (Section 4.2.4). Each subsection begins with a brief description of the overall scope of research and development topics for that topic category, along with examples of over-arching issues that need resolution, available resources and information for use in conducting the needed R&D, and examples and details of needed R&D outcomes. The one-page priority R&D topic summaries follow; 3-to-5 topics are provided for each topic category (subsection).

4.2.1. Identification and Characterization of Wind Hazards

All types of wind events are included under this category, and each type of wind hazard must be defined and delineated so that both present and future conditions can be considered. Users should be provided with the ability to determine needed parameters for wind resistant design, considering related phenomena and attributes, such as windborne debris, wind-driven rain, non-synoptic versus synoptic wind events and event probabilities. The magnitude of the hazard should be determined with variations in exposure, topography, urban interface (intersection of exposure variations), and land-sea interface (at the water line where both wind and storm surge transition). The risk of various magnitude wind events should be delineated clearly. Methods to modify the risk basis should also be available so that designs for less or greater risk events can easily be developed. The design risk basis should correspond to the design risk basis of other hazards.

Following is a summary of the current availability of sources and types of wind hazard information:

1. The wind speeds used for design are currently available in two formats: (1) in printed format in the ASCE/SEI 7-10 engineering standard, *Minimum Design Loads For Buildings and Other Structures* (ASCE/SEI, 2010); and (2) on the internet in digital form at the Applied Technology Council Windspeed Determination website (www.atcouncil.org/windspeed), which provides location-specific wind speed to the nearest 1 mph, computed from digital information based on the printed ASCE/SEI 7-10 maps. Wind speeds, which are "illustrated" by lines on the ASCE/SEI 7-10 maps along the hurricane-prone coastline, have been developed using computer simulations of hurricane wind fields making landfall, with wind speeds decreasing as the hurricane decays inland. Codification of the wind speeds has led to a simplified compartmentalization of wind speeds into three regions: the western United States (California, Oregon, and Washington), the central United States, and the hurricane coast (both the Gulf of Mexico coast and the Atlantic Ocean coast).
2. There are 1004 Automated Surface Observing System (ASOS) sites across the country operated by the National Weather Service. These stations collect and record wind speeds 24 hours per day, every day of the year. These data are archived and can be accessed for wind speed histories. No attempt to date has been made to use these data for building designs, except on a case-by-case basis. The data have not been mapped. NIST is currently working on mapping this wind speed data so that it could possibly be used in ASCE/SEI 7 and be available to practitioners.
3. There have been minimal attempts to determine a joint probability of a hurricane wind and associated storm surge; these events are mapped independently and the design event is defined by two very different mean recurrence intervals.
4. There have been minimal attempts to characterize non-synoptic wind systems or their probability.
5. There has been no attempt to define the joint probability between non-synoptic and hurricane wind systems.

The required types and levels of information needed about the wind hazard that were described by workshop participants and interpreted from the literature search are:

1. With greater spatial information available, simplified regionalization should be replaced with information about the wind hazard defined by isotachs of wind speeds or location-specific wind speed points, such that interpolation could be used everywhere in the country.
2. Given the difficulty with resolution on existing printed wind speed maps that are used for design (e.g., the ASCE/SEI 7 maps), wind speed data, as it is improved, needs to continue to be made available in digital form through the internet (as is currently done by the Applied Technology Council through the internet site, www.ATCouncil.org/ windspeed).
3. A method is needed for delineating the wind-borne debris on a regional basis. Such wind-borne debris regions should further be defined by describing the type of debris

and the level of damage that might occur from this debris, including the wind speed required to get the debris airborne.

4. Either through mapping or some other designation, the wind hazard most likely to be the design event, whether it be synoptic or non-synoptic, must be described with sufficient detail for the designer to know which design approach and criteria for wind speeds and resultant forces should be followed for a given building site.
5. For hurricane coastlines, designers should have an event defined by one mean recurrence interval that incorporates impacts from the joint high wind and storm surge hazard.

Following are the five priority research and development topics that fall under this wind hazard identification and characterization category.

R&D Priority: Wind Hazards

Topic No. 1: Hazard maps for high intensity non-synoptic events, including tornadoes

Description: Current wind hazard maps used for design of buildings do not consider all extreme wind hazards, and the unique characteristics associated with different windstorm types, such as tornadoes and downbursts. Maps and databases that provide design level information for all types of damaging windstorms are needed.

Estimated Cost: $1,000,000 to $3,000,000

Estimated Time: 5-8 years

Measurement Science Challenges and Potential Solutions

Challenges

- Tornado event history not generally captured by weather stations
- Tornado wind speeds inferred from observed damage and Enhanced Fujita (EF) scale ratings
- Identification of event type from wind climate data

Potential Solutions

✓ Algorithms to automatically identify and extract specific storm types from wind climate records

Stakeholders and Roles	
Government	Provide support/funding; conduct research; collect and archive data
Industry	Develop and critique maps; provide feedback on usefulness
Standards Organizations	Adopt maps into standards; provide feedback on quality
Building Code Organizations	Adopt maps into codes

Impacts

- Improved safety through inclusion of all significant wind hazards in the design process

R&D Priority: Wind Hazards

Topic No. 2: Effects of climate change on extreme wind event characteristics

Description: Climate change is not currently considered in the estimation of wind hazards, and the effects of climate change on the probabilities and characteristics of extreme wind events are uncertain. Research is required to determine how and if climate change affects the wind climate; subsequently, methods should be developed to account for this uncertainty for

use in development of wind hazard maps for design.
Estimated Cost: $1,000,000 to $3,000,000
Estimated Time: 2-5 years
Measurement Science Challenges and Potential Solutions

Challenges

- Lack of consensus regarding climate change impacts on characteristics of severe storms
- Validation of climate models
- Must determine if climate change could affect all types of wind events (i.e. hurricanes, tornadoes, straightline)
- Some areas of the country have very limited historical data requiring more modeling of events

Potential Solutions

✓ Investigate the range of modeling results to quantify uncertainty, and how to incorporate in wind hazard estimation

Stakeholders and Roles

Government	Provide support/funding; conduct research on future conditions.
Universities	Conduct research on the effect of climate change on wind climate
Industry	Critique maps; provide feedback on usefulness
Standards Organizations	Adopt maps into practice; provide feedback on quality

Impacts

- Improved safety margins against future changes in extreme wind climate, and increased protection against current wind hazards

R&D Priority: Wind Hazards

Topic No. 3: Quantification of wind-borne debris hazards

Description: There is a limited technical basis for the wind-borne debris hazard provisions in current building codes and standards, which rely on simple deterministic assumptions of a typical missile triggered by a single wind speed. Probabilistic methods are needed to enable site-specific and storm-type specific characterization of likely debris types, weights, and speeds.
Estimated Cost: $500,000 to $1,000,000
Estimated Time: 2-5 years
Measurement Science Challenges and Potential Solutions

Challenges

- Wind-borne debris type, size and volume dependent on many variables, including wind characteristics, sources, and accumulated damage to buildings
- Unknown vertical wind characteristics in tornadoes

Potential Solutions

✓ Numerous studies of numerical and wind tunnel experiments on debris flight characteristics

✓ Post-storm investigations to validate debris models

✓ New wind tunnel facilities in industry and academia to provide opportunities for large/full scale testing of debris flight/impact/generation

Stakeholders and Roles

Government	Provide support/funding; conduct research on wind-borne debris
Industry	Critique maps; provide feedback on usefulness
Standards Organizations	Adopt maps into practice; provide feedback on quality
Trade Organizations (wood, plywood, new materials, shutter and shelter industries)	Provide information about materials and material behavior that is important in causing debris

Impacts

- Realistic characterization of wind-borne debris hazards can lead to development of improved test methods, wind-borne debris resistant products, and a better-defined field of expected wind-borne missile impacts

R&D Priority: Wind Hazards

Topic No. 4: Hazard modification by terrain, topography, and nearby buildings

Description: Current methods for determination of the effects of terrain on local wind characteristics are cumbersome and prone to error, particularly at building sites with nonuniform surrounding terrain. Tools are needed to automate the process by taking advantage of widely available digital aerial photographic and elevation data.

Estimated Cost: $500,000 to $1,000,000

Estimated Time: 2-5 years

Measurement Science Challenges and Potential Solutions

Challenges

- Automated evaluation of surface roughness
- Estimation of topographic effects in complex terrain

Potential Solutions

- ✓ Algorithms to use web GIS-based aerial photographs and digital elevation models (e.g., Google Earth)
- ✓ Inclusion of terrain determination on windspeed web site so more wind design information is in one location

Stakeholders and Roles

Government	Provide support/funding; conduct research
Universities/Research Organizations	Conduct research; develop mapping or mapping modification techniques for terrain
Industry	Develop methodologies; critique maps or methods; provide feedback on usefulness
Standards Organizations	Adopt maps or methods into practice; provide feedback on quality

Impacts

- Improved accuracy of site-specific wind characteristics
- Improved designs of structures located on steep terrain, especially towers, signs, and buildings on bluffs, resulting in reducing losses to these structures

R&D Priority: Wind Hazards

Topic No. 5: Joint description for hurricane wind, storm surge, and wave hazards

Description: Coastal buildings and infrastructure are exposed to potentially simultaneous actions from extreme wind, storm surge, and wave hazards during hurricanes and other coastal storms, yet this loading condition is not directly addressed in codes, standards, and practice. Methods are needed to produce joint descriptions of these hazards, along with data/maps for site-specific design guidance.

Estimated Cost: $1,000,000 to $3,000,000

Estimated Time: 2-5 years (Topic 22 [Characterization of the coastal inundation hazard for low probability events] must be completed before this topic can be completed)

Measurement Science Challenges and Potential Solutions

Challenges

- Data requirements , including needed extensive libraries of high resolution hurricane wind/surge/wave simulations
- Efficient tools for data processing
- Changing conditions, including coastlines and bathymetry, sea level rise, and effects of climate change on storms

Potential Solutions

✓ Proof of concept for method developed by NIST

✓ Newly available high resolution coastal storm modeling and storm model results database systems (e.g., CSTORM-MS)

Stakeholders and Roles	
Government	Provide support/funding; conduct research
Universities/Research Organizations	Conduct research; develop mapping techniques
Industry (including the Insurance Industry)	Critique maps
Standards Organizations	Adopt maps/mean recurrence intervals into practice; provide feedback on quality

Impacts

- Risk-consistent design across the range of coastal hazards
- Improved community resilience to multiple hazards

4.2.2. Wind Loads and Effects

Wind loads on buildings and other structures are currently determined by either calculating loads using formulas and methodologies available in ASCE/SEI 7-10 (ASCE/SEI, 2010) or by using the results from wind tunnel studies. There is evidence that the current methodologies in ASCE/SEI 7-10 are so complex (or confusing) that for any particular wind load determination, no two practitioners are likely to get the same calculated result. There are many uncertainties in wind load determination, given the variability of the wind speed and changing direction, and the influence of topography, terrain and building shape. However, it seems clear that how these uncertainties are dealt with has not adequately been explained in either ASCE/SEI 7 or other documentation.

Following is a summary of the current availability of sources and types of information on wind loads and effects:

1. Wind loading methodologies have grown increasingly cumbersome to the average practitioner. There are two formula-based methods and two tabular methods for

finding wind loads on buildings. The building type and structural frame restrictions placed on each of the methods are not well understood.
2. Wind loads are determined irrespective of the type of windstorm event except there is no defined method for determining loads from tornadoes (although this will likely be addressed for the first time in the 2016 version of ASCE/SEI 7 commentary).
3. Most information on pressure coefficients related to building shape is more than 30 years old and needs to be updated in a consistent manner.

The required level of information needed about wind loading and its effects that was described by workshop participants and interpreted from the literature search is:

1. Wind loading simulation tools (e.g., computational fluid dynamics) need to be developed for use in the practice of wind engineering, and must be validated by other wind pressure testing or calculation methods.
2. Wind loading methods need to be researched for non-synoptic wind events and differences in the loads created by these events described in current engineering standards.
3. Wind loading created by internal pressurization of a "breached building envelope" caused by wind-borne debris should be modified, perhaps by requiring designs for internal pressurization unless debris impact protection is provided.
4. A more seamless transition in pressure coefficients between components and cladding (C&C) loading and main wind force resisting system (MWFRS) loading is needed.

Following are the four priority research and development topics that fall under this wind loads and effects category.

R&D Priority: Wind Loads and Effects
Topic No. 6: Pressure coefficients for wind load determination
Description: Much of the pressure coefficient data in current wind loading standards and codes is based on wind tunnel test data for rectangular cubic shapes. Pressure coefficients and wind loading procedures based on appropriate test data that address the range of actual building geometries are needed.
Program Element: 1: Provide technical support for windstorm and coastal inundation engineering practice and code development process.
Estimated Cost: > $3,000,000
Estimated Time: > 8 years
Measurement Science Challenges and Potential Solutions

Challenges

- Comparison of wind tunnel data from different labs using a consistent basis
- Pressure coefficients for use with non-synoptic winds are not well defined
- Narrowing the range of shapes and sizes to be tested

Potential Solutions

✓ Development of extensive wind tunnel databases in industry and academia for a wide range of building shapes, and methods for their effective use

Stakeholders and Roles	
Government	Provide support/funding; conduct research
Universities/Research Organizations	Conduct research; develop revised pressure coefficients and methods for their effective use
Industry	Critique revised coefficients; provide feedback on the effect revised coefficients have on design; use wind tunnels to confirm revised coefficients
Standards Organizations	Adopt changes to coefficients into practice; provide feedback on quality

Impacts

- Improved safety or efficiency of design
- Reduction of possible errors when developing pressures for building shapes not currently covered by the existing standard

R&D Priority: Wind Loads and Effects

Topic No. 7: Computational fluid dynamics tools for practitioners

Description: While computational fluid dynamics tools are commonly used in many areas of engineering practice, no such validated tools exist for wind engineering applications. Computational fluid dynamics software systems and means for verification of model results are needed.

Estimated Cost: > $3,000,000

Estimated Time: > 8 years

Measurement Science Challenges and Potential Solutions

Challenges

- Techniques to reduce computational resources needed
- Simulation of peak pressures
- Validation process and criteria

Potential Solutions

✓ Adaptation of existing open source computational fluid dynamics software for wind engineering applications

✓ Simplification of turbulence modeling to remove low frequency components

Stakeholders and Roles	
Government	Conduct research, provide funding
Universities/Research Organizations	Conduct research; develop an improved method
Industry	Validate the calculation methods with real problems and wind tunnel testing
Standards Organizations	Adopt changes in methods into practice; provide feedback on quality of results

Impacts

- Much faster, easier, less expensive, and more flexible analysis of wind loads compared to existing wind tunnel test procedures
- If coupled with building information modeling (BIM), improved efficiency in building design because loads defined by wind effects could be used to determine member sizes in one step

R&D Priority: Wind Loads and Effects

Topic No. 8: Rational approach for wind load analysis

Description: The numerous procedures for determining wind loads in current codes and standards, which are based on different datasets, methods, and assumptions, are very confusing to the engineering profession, and can provide significantly different results for a single building. A new approach is needed that provides consistent and accurate results for all building geometries.

Estimated Cost: $500,000 to $1,000,000

Estimated Time: 2-5 years (Completion of Topic 6 [Pressure coefficients for wind loads] first would improve the results of the effort spent on this topic)

Measurement Science Challenges and Potential Solutions

Challenges

- Limitations of existing pressure coefficient datasets (see Topic 6)
- Covering buildings with static and dynamic response in a single method

Potential Solutions

✓ Use of database-assisted design tools to validate potential methods

Stakeholders and Roles

Universities/Research Organizations	Conduct research; develop an improved singular method
Industry	Work with universities/research organizations to develop singular method; test the calculation methods with real problems
Standards Organizations	Adopt changes in method into practice; provide feedback on quality of method

Impacts

- Increased consistency in repeatable results and reduced errors associated with computation of wind loads

R&D Priority: Wind Loads and Effects

Topic No. 9: Modeling and simulation for determination of wind loads due to non-synoptic winds, including tornadoes

Description: There is limited information about the loading effects on buildings from non-synoptic winds, and building occupants are in danger of death or injury from these types of events, in addition to significant property damage occurring annually. Additional modeling or simulation of the wind effects from these events needs to be conducted so wind load standards can be developed to be used by practitioners.

Estimated Cost: > $3,000,000

Estimated Time: > 8 years

Measurement Science Challenges and Potential Solutions

Challenges

- Lack of data on non-synoptic wind hazard and load characteristics

Potential Solutions

✓ new tornado simulators at several institutions

Stakeholders and Roles

Government	Provide support/funding; conduct research
Universities/Research Organizations	Conduct research; develop revised pressure coefficients

Industry	Critique revised coefficients; provide feedback on the effect revised coefficients have on design; use wind tunnels to confirm revised coefficients
Standards Organizations	Adopt changes to coefficients into practice; provide feedback on quality

Impacts

- Performance improvements for buildings subject to non-synoptic winds that would save lives, reduce injuries, and reduce damage
- Improved community resiliency to recover faster from severe non-synoptic wind events

4.2.3. Wind Resistance

Resistance is determined either by calculation of the strength of materials and connections or by test methods developed by independent test organizations and employed by a product manufacturer or product trade organization. Buildings have many components and connections, such that the cause of failure of a building due to high winds is often difficult to determine. The failure could be initiated by either component/connection failure or structural frame movement during the event and eventual failure from cyclic loading or from a breach of the building envelope, leading to a change in the aerodynamic characteristics. Knowledge of the actual level of safety built into any wind design remains uncertain except for the testing performed by product manufacturers and trade organizations, which is component- based rather than system-based. However, most testing is done statically, and not dynamically, so the testing that is done does not closely represent an actual wind event. There are an extremely large number of buildings already built along hurricane-prone coastlines, and in tornado-prone areas, that might need to be retrofitted to yield better wind performance, yet currently there is little known about what level of resistance to a high-wind event actually exists for those buildings. Thus, it is not possible to determine how much retrofitting would be appropriate for such buildings. To further complicate the problem of determining wind resistance, there has been little work done on identifying the reduction in resistance as buildings get older and deterioration in both materials and connections occurs.

Following is a summary of the current availability of sources and types of information on wind resistance:

1. Resistance of materials is determined by materials testing of some limited number of products. Most test methods provide static resistance.
2. Capacities of mechanical connectors are given in allowable loads; therefore, some unknown factor of safety has been used to reduce ultimate loads determined by tests to the allowable load.
3. Capacities of members may be provided in either allowable or ultimate stresses, depending on the material type. Allowable stresses account for some unknown factor of safety used to reduce ultimate failure test results to allowable stresses.
4. There is no current methodology available to reduce expected resistance of either materials or connections based on age, weathering, deterioration (from weathering or from material-destroying organisms) or other age-related strength reductions.

The required level of information needed that was described by workshop participants and interpreted from the literature search is:

1. Test methods need to be developed to realistically determine the wind resistance of components and cladding materials in installed conditions for materials, aging, wind-driven rain, water infiltration, material deterioration, and other conditions, such as those caused by heat, thermal expansion and contraction, and corrosion.
2. Guidance on wind retrofitting of residential buildings is newly available (FEMA, 2013) ; however, this guidance assumes effectiveness of the entire structure once the retrofit is complete as though the rest of the structure was new, which is not the case. Important differences in performance of the retrofit measures are expected, since the structures being retrofitted are existing. These differences need to be studied and understood so adjustments in the retrofit techniques can be made as needed to achieve optimum performance.
3. Test methods for materials should be developed using a dynamic analysis technique to better replicate the cyclic nature of the wind loading (dynamic load testing for glass does exist and some test methods have been developed in other parts of the world).
4. Practitioners need better information about the expected levels of safety factors used for the products and materials they are specifying and using for design.
5. Practitioners need information about system-level safety factors that require analytical tools to assess.

Following are the five priority research and development topics that fall under this wind resistance category.

R&D Priority: Wind Resistance
Topic No. 10: Test methods for resistance to cyclic loading
Description: Existing test methods for wind pressure resistance to cyclic loading do not reflect realistic loading conditions. New methods are needed that are verified to produce effects comparable to those experienced in service.
Estimated Cost: $1,000,000 to $3,000,000
Estimated Time: > 8 years
Measurement Science Challenges and Potential Solutions

Challenges

- The wide range of building components and materials and thus the need to select some strategic number of components and materials for testing improvements
- Connection of tested component to building must also be accounted for

Potential Solutions

✓ New wind tunnel/wind load experimental facilities in industry and academia to provide opportunities for large/full scale testing to validate effectiveness of proposed test methods

Stakeholders and Roles

Government	Provide sport/funding; conduct research
Universities/Research Organizations	Conduct research; develop test equipment and test protocols

Industry	Test and use new testing apparatus; provide results to user groups and practitioners; compare results to calculated requirements
Standards Organizations	Adopt changes to test methods into standards of practice; provide feedback on quality of developed test methods

Impacts

- Improved component and cladding products and their attachment schemes to the building frame
- Reduced wind-borne debris fields

R&D Priority: Wind Resistance

Topic No. 11: Field measurement techniques and test methods to evaluate condition of existing construction

Description: There is a lack of methods and tools for field evaluation of the condition of installed wall and roof cladding elements. New measurement techniques and methods are needed that provide information on the strength, condition, remaining service life, and ability to resist penetration of water and wind-driven rain.

Estimated Cost: > $3,000,000

Estimated Time: > 8 years

Measurement Science Challenges and Potential Solutions

Challenges

- Likely require new test and measurement equipment
- Wide range of building cladding types and materials
- Attachment of cladding to building is often not exposed or easily accessible

Potential Solutions

✓ Adaptation of nondestructive evaluation testing methods for structural materials

Stakeholders and Roles

Government	Provide support/funding; conduct research
Universities/Research Organizations	Conduct research; develop test equipment and test protocols
Industry	Test and use new testing apparatus; provide results to user groups and practitioners; compare results to calculated requirements
Standards Organizations	Adopt changes to test methods into standards of practice, provide feedback on quality of developed test methods

Impacts

- More robust windstorm vulnerability assessments and early warning of problems before evidenced through failures
- Improved assessment of community resiliency and potential decreased service downtime in affected buildings

R&D Priority: Wind Resistance

Topic No. 12: Explicit factors of safety for materials and assemblies

Description: As wind and flood design progresses to ultimate strength and becomes performance-based, knowledge about the stress limit states of both building frame and component materials and assemblies becomes more critical. While material development work is conducted by trade groups, there is no limit-state development work being conducted on assemblies or systems. Knowledge about the levels of safety in design is crucial to preventing failures in the future.

Estimated Cost: $500,000 to $1,000,000

Estimated Time: 2-5 years

Measurement Science Challenges and Potential Solutions

Challenges

- Large body of materials and systems

Potential Solutions

✓ Modeling or simulation of material behavior grouped into a system or assembly with predictive stresses at failure

Stakeholders and Roles	
Government	Provide support/funding; conduct research
Universities/Research Organizations	Conduct research; develop test equipment and test protocols
Industry	Test materials-components-systems; provide results to user groups and practitioners; compare results to results obtained with use of ultimate wind speeds
Standards Organizations	Adopt changes to test methods into standards of practice, provide feedback on quality of developed test methods

Impacts

- Improved building system reliability and improved information for performance-based design

R&D Priority: Wind Resistance

Topic No. 13: Methods or models to predict loss of strength due to aging and deterioration

Description: Field investigations after high wind and coastal inundation events often indicate failures in materials and in structural and component connections due to aging or deterioration, yet there is no existing method to measure the extent of reduced resistance from these factors. A method for estimating the possible loss of strength of original construction due to these factors is needed. In addition, retrofit techniques need to be developed to counteract the lack of or loss of this resistance in existing buildings.

Estimated Cost: $500,000 to $1,000,000

Estimated Time: 2-5 years

Measurement Science Challenges and Potential Solutions

Challenges	*Potential Solutions*
• No current method or standard of practice to account for aging. • Numerous building materials and building systems to consider.	✓ Develop proxy method for loss of resistance based on existing material properties (e.g., moisture content, reduced section of steel from corrosion)

Stakeholders and Roles

Government	Provide support/funding; conduct research
Universities/Research Organizations	Conduct research; develop test equipment and test protocols
Industry	Develop retrofit techniques using materials not currently used for retrofitting; conduct outreach and training on retrofit techniques; provide results to user groups and practitioners, compare results to current methods

Impacts

- Significantly reduced damage from windstorm events
- Improved performance for components and cladding (where most wind losses occur)

R&D Priority: Wind Resistance

Topic No. 14: Guidelines for wind vulnerability assessment of existing buildings

Description: There is a lack of guidance available for assessment of the vulnerability of existing buildings to windstorms. The limited existing guidance is primarily related to critical facilities and is qualitative in nature. Quantitative methods of assessment are needed that are more generally applicable.

Estimated Cost: $500,000 to $1,000,000

Estimated Time: 2-5 years

Measurement Science Challenges and Potential Solutions

Challenges	*Potential Solutions*
• Data to support quantifiable assessments • Limited methods for evaluation of condition of installed cladding (see Topic 11)	✓ Adaptation of existing methods, guidelines and standards for seismic capacity and vulnerability evaluation

Stakeholders and Roles

Government	Provide support/funding
Universities/Research Organizations	Research; develop test equipment and test protocols
Industry	Develop guidance based on available research and data, including new testing apparatus and protocols; provide results and training to practitioners and other user groups; compare evaluation results to observed high wind effects
Standards Organizations	Convert guidelines into standards of practice

Impacts

- Improved measurement of building and community resiliency through a uniform basis to evaluate vulnerability
- Improved understanding of needed mitigation actions to address identified vulnerabilities

4.2.4. Performance-Based Engineering Methodology

Performance-based design for wind engineering should cover wind hazards for hurricanes, extra-tropical storms, and non-synoptic winds such as thunderstorms and tornadoes. The use of performance-based design analysis methods are intended to provide a more direct connection between the design of the building and the owner- expected and required performance of that building. Specifying a specific level of performance for a defined hazard level, and quantifying the consequences to the owner for such a hazard could improve understanding of building design and operation and aid in developing more optimum designs to achieve the desired performance. The current static analysis methods provide conservative results and do not consider post-elastic behavior and stresses beyond the yield strength. The method used for performance-based design of wood, steel, and concrete structures will vary somewhat, depending on material properties.

Following is a summary of the current availability of sources and types of information on performance-based engineering for wind design:

1. There is little written information about performance-based wind engineering that has made its way into practitioner use.
2. There are examples of limited uses of the concept on a project-by-project basis. The examples show promise in taking advantage of reserve structural capacity in order to design more cost-effective structures.
3. Performance-based design procedures have been developed for the seismic hazard; the application of performance-based design methods for wind is expected to generally be based on the approaches used in seismic design.

The required level of information needed that was described by workshop participants and interpreted from the literature search is:

1. Performance targets and acceptability criteria need to be developed for use in wind engineering
2. A wind engineering performance-based design approach should be developed that is analogous to the ASCE 41 standard, *Seismic Rehabilitation of Existing Buildings*, which is based on extensive FEMA-funded guidance development efforts (an important step preceding the adoption of standards).
3. Tools are needed to define the hazard and to conduct assessments of both new and existing buildings.
4. Performance-based designs performed across the spectrum of a community should be linked to resiliency and sustainability.

Following are the four priority research and development topics that fall under this wind engineering performance-based design category.

R&D Priority: Wind Performance-Based Design

Topic No. 15: Performance levels and acceptance criteria for wind hazards

Description: There is a lack of tools and guidance available to design and evaluate buildings to withstand windstorms on a performance basis. A critical first step towards enabling performance-based design is to define expected levels of performance and acceptance criteria for different hazard levels.

Estimated Cost: $1,000,000 to $3,000,000

Estimated Time: > 8 years

Measurement Science Challenges and Potential Solutions

Challenges	*Potential Solutions*
• Requires integration of technical and societal/policy-level input	✓ Adaptation of performance-based seismic design and evaluation guidance

Stakeholders and Roles

Government	Provide support/funding
Universities/Research Organizations	Develop and refine performance-based design methodologies
Industry	Develop guidelines for practitioners; conduct outreach and training for practitioners
Standards Organizations	Convert guidelines into standards of practice

Impacts

- Cost-effective performance-based design procedures that, when implemented, result in substantial improvement in building performance during extreme wind events
- The adoption of performance-based seismic design procedures in codes and standards provide the potential for understanding possible reduction in wind hazard losses

R&D Priority: Wind Performance-Based Design

Topic No. 16: Analysis procedures for nonlinear system behavior used in performance-based design

Description: Current design procedures for resisting wind loads assume linear elastic material behavior. There is a lack of procedures for applying nonlinear analysis methods for wind loads. Such methods are needed to provide the ability to predict building performance to failure.

Estimated Cost: $500,000 to $1,000,000

Estimated Time: 5-8 years

Measurement Science Challenges and Potential Solutions

Challenges	*Potential Solutions*
• Procedures are needed for both strength and serviceability limit states • Many types of building envelope materials and building systems	✓ Adaptation of non-linear analysis methods used in seismic design and in design to avoid progressive collapse

Stakeholders and Roles

Government	Provide support/funding

Universities/Research Organizations	Develop analysis procedures
Industry	Develop guidelines for practitioners that incorporate the analysis procedures; conduct outreach and training for practitioners
Standards Organizations	Convert guidelines into standards of practice

Impacts

- Improved methods for analysis of building performance for rare, extreme events
- Material and cost savings in buildings designed using these analysis procedures

R&D Priority: Wind Performance-Based Design

Topic No. 17: Cyber-based tools to support performance-based design

Description: Engineering, modeling and simulation tools used in performance-based design should be broadly available to the profession electronically so the various tools can be tested and used. Broadly available tools and information will speed up the rate at which performance-based design is accepted as part of normal practice. These tools might include those used for structural analysis, modeling wind behavior on a building, determining serviceability limits and effects, modeling terrain effects on buildings and similar methods for studying wind effects on buildings.

Estimated Cost: $1,000,000 to $3,000,000

Estimated Time: 5-8 years

Measurement Science Challenges and Potential Solutions

Challenges

- Some proprietary issues may slow down offering tools to the public
- Accumulation of information and task of determining how to make it available are not inconsequential

Potential Solutions

✓ Tools and information that flow from NIST so that most tools will be in the public domain

Stakeholders and Roles

Government	Provide support/funding; conduct research and develop cyber tools
Universities/Research Organizations	Conduct research; develop cyber tools and metrics
Industry	Test cyber tools; develop guidance for their use; provide results to user groups and practitioners; compare evaluation results
Standards Organizations	Convert guidance into standards of practice, if necessary

Impacts

- More rapid delivery of information to the practice on performance-based design

R&D Priority: Wind Performance-Based Design

Topic No. 18: Measurement of windstorm resilience and benefits of performance-based design

Description: Measures of community resilience specific to windstorm hazards, and documentation on associated benefits, are lacking and are in need of development.

Estimated Cost: $500,000 to $1,000,000

Estimated Time: 5-8 years

Measurement Science Challenges and Potential Solutions

Challenges

- Measurement of resilience in general is still ill-defined
- Link between performance-based design and resiliency has not previously been attempted

Potential Solutions

✓ Leveraging NIST's ongoing efforts to develop methods for measurement of resilience

Stakeholders and Roles

Government	Provide support/funding, develop resiliency initiatives
Universities/Research Organizations	Conduct research; develop performance-based design metrics specific to community resiliency
Industry	Evaluate performance-based design metrics; develop guidance to apply metrics
Standards Organizations	Convert guidelines into standards of practice, as necessary

Impacts

- More rapid delineation of the positive benefits of performance-based design, leading to more rapid adoption of performance-based design by the design profession

4.3. Coastal Inundation Priorities

The priority coastal inundation R&D topics are elucidated in the following four subsections: identification and characterization of coastal inundation hazards (Section 4.3.1); coastal inundation loads and effects (Section 4.3.2); resistance to coastal inundation (Section 4.3.3); and performance-based engineering design for coastal inundation (Section 4.3.4). As in the case for wind, each subsection begins with a brief description of the overall scope of research and development topics for that topic category, along with examples of over-arching issues that need resolution, available resources and information for use in conducting the needed R&D, and examples and details of needed R&D outcomes. The one-page priority R&D topic summaries follow; 2-to-4 topics are provided for each topic category (subsection).

4.3.1. Identification and Characterization of Coastal Inundation Hazards

The coastal inundation hazard must include the effects of future conditions (e.g., climate change and future development), must consider inland effects in bays and tributaries that are sometimes great distances from tidal-influenced coastlines, should be characterized by volume and type of flood-borne debris, must be correlated with the wind influence on the extent of inundation, and must consider the influence of the slope and roughness of the coastal floor and the channeling effects of the onshore built environment.

Following is a summary of the current availability of sources and types of information on coastal inundation hazards:

1. Current hazard identification is provided by two primary sources: (1) community specific flood maps that delineate the hazard by the limit of inundation for various mean recurrence interval storms; and (2) storm surge inundation limits from a

maximum of maximum flood events illustrated on maps used for evacuation purposes.

2. Flood maps published by FEMA are intended for insurance rating purposes primarily; they are used for design because they are the only design tool with inundation extent indicated, and water surface elevation expectations noted, that are based on the probability of a specific flood elevation occurring.
3. Flood-borne debris is not characterized at all for any coastal site.
4. There is no inundation depth information on flood maps although new mapping graphical interfaces are beginning to provide site-specific depths in some limited number of communities across the nation.
5. Some coastal communities have very recent flood hazard information and some are very outdated. The revisions and updating only occur when the community, the state, or FEMA provide funding to update the flood maps.
6. Tsunami hazard maps are mainly produced for evacuation and do not contain information needed for engineering design (until a new tsunami loads section is added to ASCE/SEI 7 in 2016).

The required level of information needed about the coastal inundation hazard that was described by workshop participants and interpreted from the literature search is:

1. Hazard identification information must include ways to consider future conditions such as sea-level rise, long-term erosion, subsidence, and future development.
2. There should be a flood-borne debris data layer for flood mapping that would help determine the possible magnitude and extent of this hazard.
3. Tsunamis should be included in the coastal inundation hazard mapping.
4. Methods to determine the hazard for lower-probability but more extreme events (lower than 1% and 0.2% annual probability) should be developed.
5. Wave heights need to be provided for lower-probability events.
6. Hazard information should be provided in a data repository easily accessed by those who need this information.
7. Ways to define and describe the hazard other than just water depth and water inundation limits should be explored. These definitions and descriptions should separate engineering information needs from insurance information needs, given that the flood maps were, and still are, used primarily for flood insurance rating for individual properties located in floodplains.

Following are the four priority research and development topics that fall under this coastal inundation hazard category.

R&D Priority: Coastal Inundation Hazards

Topic No. 19: Effects of sea level rise and future shoreline changes on coastal inundation hazards

Description: Current methods for defining the coastal flood hazard ignore future conditions such as sea level rise and shoreline changes. Methods must be developed to characterize the coastal inundation hazard for future conditions such as sea-level rise, subsidence, long-term erosion, and future development, perhaps by specifying scenarios, including the number of

years in the future that are being considered. This must be done at a community scale since broader implications of sea level rise are beyond the scope of this roadmap. This flood hazard characterization should be consistent with wind speed hazard characterizations.
Estimated Cost: $500,000 to $1,000,000
Estimated Time: 5-8 years
Measurement Science Challenges and Potential Solutions

Challenges

- Creating future conditions as a data layer
- Estimating sea level rise for site-specific locations

Potential Solutions

✓ Adapt information from Intergovernmental Panel on Climate Change to create scenarios based on a rate of rise times a specified number of years into the future

Stakeholders and Roles:

Government	Conduct research on possible future conditions; identify historical references to use as a basis to predict future conditions
Universities/Research Organizations	Conduct research; provide historical references to use as a basis to predict future conditions
Industry	Create scenario visualization tools for use by regulators and design practitioners

Impacts

- Scenario development for future conditions that would improve decision making about foundation types, depth of embedment of pile foundations, and siting approaches
- Visualization tools demonstrating scenarios that would illustrate the impact of future conditions to communities

R&D Priority: Coastal Inundation Hazards
Topic No. 20: Quantification of flood-borne debris hazards
Description: Probabilistic methods are needed that can estimate debris generation and debris characteristics based on flood variables (depth, waves, velocity) and then a methodology needs to be developed for determining how flood-borne debris 'layers' could be used with flood mapping technology to describe the possible magnitude (quantity, size, weight and speed) and extent (area of flooded surface) of debris, so flood loads from debris could be determined from this information.
Estimated Cost: $1,000,000 to $3,000,000
Estimated Time: 5-8 years (Topics 21 [Effects of over-land flow] and 26 [Flood damage functions] must be completed before this topic can be completed)
Measurement Science Challenges and Potential Solutions

Challenges

- Using satellite tools to predict magnitude of debris
- Mapping inland extent of debris and combining with other flood hazard data

Potential Solutions

✓ Create an overlay tool that works with a geospatial image that would estimate debris within some range of distances from a site and that could be used to define flood debris characteristics

Stakeholders and Roles:	
Government	Provide technical support and possible access to other geospatial tools that could accelerate research and development on this topic
Universities/Research Organizations	Conduct research; develop methods for aggregating data meaningfully for flood-borne debris
Industry	Test and refine graphics tools

Impacts

- Specific data about both the extent and magnitude of flood-borne debris that would improve decision making about siting approaches and the magnitude of the debris load at any particular building location
- Visualization tools demonstrating flood-borne debris scenarios that would illustrate to communities the impact on various design decisions

R&D Priority: Coastal Inundation Hazards

Topic No. 21: Effects of over-land flow on waves and other coastal inundation hazards

Description: The effects of over-land flow, including waves, debris, flood velocity, wind-driven influences, erosion effects at buildings and channeling effects of the built environment need to be determined. Ways to characterize these effects into a comprehensive hazard identification system are also needed.

Estimated Cost: $1,000,000 to $3,000,000

Estimated Time: 5-8

Measurement Science Challenges and Potential Solutions

Challenges

- Methods to determine the extent of erosion at buildings from either inflow or outflow are not well developed or tested.
- No methodology exists for determining the effects of channeling of flow between buildings or other obstructions in the built environment

Potential Solutions

- ✓ Creating a web-based tool that determines erosion from historical data at a specific site
- ✓ Using satellite imagery, develop a method for determining channeling effects between obstructions

Stakeholders and Roles:	
Government	Provide support/funding
Universities/Research Organizations	Conduct research; develop metrics for over-land flow
Industry	Evaluate metrics for over-land flow; develop guidelines for implementation of a comprehensive hazard identification system that incorporates methods to determine the effects of overland flow
Standards Organizations	Convert guidelines into standards of practice, if necessary

Impacts

- Improved community resilience through consideration of overland flow effects
- Improved community zoning, floodplain ordinances, flood insurance rates, and building codes

R&D Priority: Coastal Inundation Hazards

Topic No. 22: Characterization of the coastal inundation hazard for low probability events

Description: Methods for estimating the full range of storm surge and tsunami coastal inundation hazards (including depth/elevation, velocity, and wave heights and periods) for low-probability events are lacking and need to be developed. Tools must then be created to map these hazard characteristics.

Estimated Cost: $1,000,000 to $3,000,000

Estimated Time: 5-8

Measurement Science Challenges and Potential Solutions

Challenges

- Each varying segment of the coastline may need to be studied for extremely long return period events
- Creating an ultimate-strength building design for a low-probability coastal inundation event that is compatible with an ultimate strength design for a corresponding wind event

Potential Solutions

✓ Creating a web-based tool that would determine return periods at a specific site using shoreline topography

Stakeholders and Roles:

Government	Provide support/funding; conduct research
Universities/Research Organizations	Conduct research; develop methods for revising the mean recurrence interval of a design event
Industry	Develop guidelines to revise mean recurrence intervals for design events, based on new research
Standards Organizations	Adopt guidelines into standards of practice, as necessary

Impacts

- Improved knowledge about community vulnerabilities to low probability coastal inundation events could improve storm surge warning systems and improve the criteria for identifying evacuation areas.
- Consistency in probability-based design for windstorm and coastal inundation

4.3.2. Coastal Inundation Loads and Effects

Loads for coastal inundation conditions are calculated from formulas that are in ASCE/SEI 7 (Section 5), the USACE *Coastal Engineering Manual*, the FEMAP-55 *Coastal Construction Manual*, and the FEMA P-646 *Guidelines for Design of Structures for Vertical Evacuation from Tsunamis*. There are formulas for loads on buildings from hydrostatic, hydrodynamic, breaking wave, and flood-borne debris conditions. The formulas in all of the above references are essentially the same. The determination of the loads requires information about flood velocity and flood depth at the site-specific location of interest. Given the lack of knowledge about flood velocity and the impact that storm forward speed, wind speed or coastal bathymetry may have on flood velocity, the current method for determining velocity is derived primarily from flood depth. There are significant uncertainties in determining flood loads, and there is not wide spread use yet of the current flood load knowledge by practitioners.

Following is a summary of the current availability of sources and types of information on coastal inundation loads and their effects:

1. Flood velocity is currently derived from flood depth at the near shore or inland locations, depending on the building site.
2. Hydrodynamic flow is a function of flood velocity.
3. Breaking wave loads are determined from wave heights, which are derived from flood depths at the building site.
4. Flood-borne debris loads are derived from flood velocity and requires an assumption about debris type and weight. All debris is assumed to float at the still water level and at the same velocity as the water.
5. There is little guidance on how individual flood loads should be combined with each other to account for the simultaneous influence of various flood aspects (e.g., the combined influence of breaking wave loads occurring at the same time as hydrodynamic loads)

The required level of information needed about coastal inundation loads and their effects described by workshop participants and interpreted from the literature search is:

1. Laboratory testing in wave tanks should be conducted to validate flood loads on buildings for waves, velocity, debris and the effects of scour.
2. Flood load combinations should be developed to account for the simultaneous influence of various flood loads that include probability and uncertainty of simultaneous occurrence.
3. Load formulas should be studied for waves breaking onto buildings above the lowest floor.

Following are the three priority research and development topics that fall under this coastal inundation loads and effects category.

R&D Priority: Coastal Inundation Loads
Topic No. 23: Field data collection of flood loads
Description: Methods of determining flood loads on buildings caused by waves, velocity, and debris, including the effects of scour, need to be validated and tested. The data required for this validation requires instrumentation that can be field deployed to capture wave heights, water velocity, water depths and wave direction. Building response to flood loads must also be determined, which requires buildings be instrumented to determine load effects.
Estimated Cost: > $3,000,000
Estimated Time: 5-8
Measurement Science Challenges and Potential Solutions

Challenges

- Measurement devices do not currently exist.
- Instrumenting existing buildings is expensive and landfalling storm locations uncertain and infrequent
- Determining storm surge information across the flood field and not just at fixed surge measurement devices

Potential Solutions

- ✓ Leveraging existing university and government storm surge data collection efforts
- ✓ Using existing tsunami test basins for storm surge load validation
- ✓ Pre-instrument locations to collect flood data

Stakeholders and Roles:

Government	Provide support/funding, develop field measurement techniques
Universities/Research Organizations	Conduct research; develop and test load validation methods, develop field measurement techniques
Industry	Develop guidance for use of validated flood loads by design practitioners
Standards Organizations	Adopt new guidance into standards of practice

Impacts

- Improved flood load accuracy for design of buildings and other structures in coastal inundation areas
- Improved information about building response to flood loads, allowing designers to develop building framing systems more capable of resisting the loads and thus reducing damage

R&D Priority: Coastal Inundation Loads

Topic No. 24: Load combinations for simultaneous flood hazards

Description: Flood load combinations need to be developed to account for the simultaneous impacts of various flood forces, such as those generated by breaking waves, moving water and flood-borne debris.

Estimated Cost: $500,000 to $1,000,000

Estimated Time: 2-5 years (Topic 22 [Characterization of the coastal inundation hazard for low probability events] and Topic 23 [Field data collection of flood loads] must be completed before this topic can be completed)

Measurement Science Challenges and Potential Solutions

Challenges

- Load combinations might depend on time sequence of greatest magnitudes of the flood components
- Flood load return periods for design must be determined before load combinations can be created

Potential Solutions

- ✓ Verifying numerical models with field data would provide confidence in developing designs more resistant to floods

Stakeholders and Roles:

Government	Provide support/funding
Universities/Research Organizations	Conduct research; develop appropriate load combinations

Industry	Develop guidelines for applying the new flood load combinations in design
Standards Organizations	Convert guidelines into standards of practice

Impacts

- Improved design and performance of buildings and other structures during floods in coastal inundation areas, resulting in reduced losses

R&D Priority: Coastal Inundation Loads

Topic No. 25: Validation of flood forces on buildings

Description: Flood load estimation procedures must be validated for coastal inundation forces impacting a building. The load information could come from field collected information, current formulas and calculation techniques, numerical models, or laboratory testing.

Estimated Cost: $1,000,000 to $3,000,000

Estimated Time: 5-8 years (Topic 23 [Field data collection of flood loads] must be completed before this topic can be completed)

Measurement Science Challenges and Potential Solutions

Challenges

- Verifying wave loads on buildings through testing and numerical modeling
- Establishing realistic boundary conditions for use in numerical models

Potential Solutions

- ✓ Using computational fluid dynamics to model interface and impact between water loads and buildings

Stakeholders and Roles:

Government	Provide support/funding
Universities/Research Organizations	Conduct research; develop appropriate equations
Industry	Develop guidelines for applying flood (wave) load calculations in the design of new and the retrofit of existing buildings
Standards Organizations	Convert guidelines into standards of practice

Impacts

- Reduced coastal flood damage in new and retrofitted buildings that can resist or deflect wave forces
- Would potentially provide a computer-based approach to study flood effects on existing buildings prior to a storm surge event.

4.3.3. Resistance to Coastal Inundation

The resistance to the effects of coastal inundation is typically accomplished by reducing the possibility of being inundated. The most effective way to resist the effects of inundation is to not be located in a coastal inundation area (i.e., stay outside of the coastal floodplain). The next most effective way to resist inundation is to be above the inundation height, whatever that height is for any storm event. Building survival is frequently accomplished if the foundation survives, and if load path continuity is maintained throughout the structural systems. There is very little engineering design of the building to resist flooding, given what is currently known about flood forces, so the solutions that are known to work are removal of the building from the floodplain or elevation of the building above the expected flood elevation.

Following is a summary of the current availability of sources and types of information on resistance to coastal inundation:

1. Elevate the building above the flood (top of the waves) to eliminate the possibility of waves striking the building structure at or above the floor system. The expected wave impact assumes that service connections will be severed and will need to be replaced.
2. Foundations, such as piles or columns, must be specifically designed to withstand flood forces and to resist other lateral and vertical loads; as a result, the foundations can survive a "design flood event".
3. Little is known about how to design the building above the top of the foundation to resist flood forces.
4. Erosion and scour effects are known to affect foundation performance because of the loss of supporting soil around the foundation, but the critical amount of soil loss is not known.
5. There is little variation in foundation materials used; thus, the benefits of new materials are never discovered, and there are few known techniques to strengthen an existing pile foundation.

The required level of information needed about the resistance to coastal inundation that was described by workshop participants and interpreted from the literature search is:

1. Flood damage functions should be studied to give designers a better idea of the extent of loss likely at various levels of damage that occur at various levels of flood depth.
2. Fragility curves should be developed for various building components, systems and infrastructure elements that reflect damage by flood forces.
3. Foundation type and depth should be studied to provide better design guidance on foundations that are required to resist the effects of flood forces, erosion and scour.
4. Building materials and systems located below the lowest floor, and thus in the path of flood forces, should be studied to develop increased resistance to damage from flooding.
5. Test methods need to be developed to test the materials and systems used in the various flood scenarios.

Following are the three priority research and development topics that fall under this resistance to coastal inundation category.

R&D Priority: Resistance to Coastal Inundation
Topic No. 26: Flood damage functions and fragility curves
Description: Flood damage functions and fragility curves should be developed to give designers better guidance on the magnitude of likely damage from various magnitude flood events. This information would also aid in determining when additional freeboard would significantly improve building performance.
Estimated Cost: $500,000 to $1,000,000

Estimated Time: 2-5 years

Measurement Science Challenges and Potential Solutions

Challenges	Potential Solutions
• Fragility curves need to be developed for a significant number of components and systems • Levels of flood damage have not been defined	✓ Laboratory testing of materials and some systems could provide significant information for fragility curve development

Stakeholders and Roles:

Government	Provide support/funding and flood damage information
Universities/Research Organizations	Conduct research; develop damage functions and fragility curves
Industry	Develop guidelines for considering flood-damage functions and fragility curves in the design process
Standards Organizations	Convert guidelines into standards of practice

Impacts

- Reduced flood damage and improved community resiliency through use of flood-damage functions (fragility curves) in the design of systems and components that undergo flooding
- Improved vulnerability and risk assessments conducted for new and existing structures, and possibly improved HAZUS loss estimates.

R&D Priority: Resistance to Coastal Inundation

Topic No. 27: Effects of erosion and scour on deep foundations

Description: Foundation design needs to be improved with a focus on pile material and spacing, embedment depths, and the load effects caused by erosion and scour at the foundation. Materials and system improvements could lead to less expensive and/or easier to install foundations that would result in reduced damage from coastal inundation events. Testing of foundation aging and the soil consolidation that occurs over time would provide very useful input into new foundation design.

Estimated Cost: $500,000 to $1,000,000

Estimated Time: 2-5 years (Topic 23 [Field data collection of flood loads] must be completed before research and development on this topic can begin)

Measurement Science Challenges and Potential Solutions

Challenges	Potential Solutions
• Soil properties vary greatly along the storm surge-prone coastlines • Limited research on the effects of scour and erosion at building foundations	✓ Soil modification techniques might reduce loss of soil ✓ New pile material could improve resistance to flood loads, minimizing impact from loss of soil

Stakeholders and Roles:

Government	Provide support/funding; conduct research
Universities/Research Organizations	Conduct research; develop improved foundation and pile design methods, including improved material specifications

Industry	Develop guidelines for improved foundation design to resist the effects of scour and erosion.
Standards Organizations	Convert guidelines into standards of practice

Impacts

- Reduced foundation damage, resulting in more rapid post-storm recovery and reduced coastal inundation impacts on communities.

R&D Priority: Resistance to Coastal Inundation

Topic No. 28: Test methods for materials and systems subject to flooding

Description: Test methods do not currently exist and must be developed for materials and systems installed to improve the performance of building components located below the Base Flood Elevation (BFE) that obstruct the flow of flood waters (e.g., breakaway walls). When such components break loose, they become flood- borne debris. New materials and systems for such components also need to be developed.

Estimated Cost: $1,000,000 to $3,000,000

Estimated Time: 5-8 years

Measurement Science Challenges and Potential Solutions

Challenges

- Federal regulations regarding floodplain development may impede acceptance of new materials or systems
- Performance testing under full flood loads

Potential Solutions

✓ Materials or systems with flow through capability

✓ Materials that might disintegrate when flooded but remain stable when dry

Stakeholders and Roles:

Government	Provide support/funding; incorporate acceptable materials/systems in current standards
Universities/Research Organizations	Conduct research; test materials and systems in flow tanks
Industry	Evaluate proposed materials and/or systems

Impacts

- Reduced flood-borne debris
- Materials and systems that comply with current flood regulations.

4.3.4. Performance-Based Design for Coastal Inundation

The National Flood Insurance Program (NFIP) language in the Code of Federal Regulations (CFR) 44, Section 60.3, is performance-based in its approach. The NFIP language says:

- "Buildings must be sited reasonably safe from flooding.
- Buildings must be designed to prevent flotation, collapse and permanent lateral movement during flooding.
- Buildings must be designed using flood-resistant materials below the Base Flood Elevation (BFE).
- Buildings must be constructed to minimize flood damage.

- The HVAC (heating, ventilation, air conditioning) and plumbing equipment must be designed to prevent water entry into those systems."

The use of performance-based design analysis methods are intended to provide a more direct connection between the design of the building and the owner-expected and required performance of that building. Specifying a specific level of performance for a defined hazard level, and quantifying the consequences to the owner for such a hazard could improve understanding of building design and operation and aid in developing more optimum designs to achieve the desired performance. However, there are no performance-based design metrics for the performance of either communities or individual buildings during a coastal inundation event. The premise behind minimizing losses during flooding has always been to "run from flood" to minimize the loss of life. Reduced building damage is expected to occur when it is elevated above the defined flood elevation (Base Flood Elevation) for the 1% annual probability event. Freeboard (additional elevation above the Base Flood Elevation) is either mandated by community flood ordinances or included as a best practice to provide a margin of safety.

Following is a summary of the current availability of sources and types of information about performance-based design, as it relates to coastal inundation:

1. Other than the performance statements in the National Flood Insurance Program language (see above), there are no performance-based design standards used in flood-resistant design.
2. The current information about magnitude of flood damages resides in the insurance claims information of the National Flood Insurance Program and lags several years behind actual events.
3. The current damage information caps losses at 50% of the building value, so any loss greater than 50% is largely undetermined from the current insurance claims data.
4. Most losses occur to residential buildings; thus, there is minimal information about non-residential building losses caused by coastal inundation.

The required level of needed information about performance-based design for coastal inundation that was described by workshop participants and interpreted from the literature search is:

5. Performance-based design information is needed for storm surge, waves, and tsunamis.
6. The engineering requirements for buildings in the flood plain need to be separated from the insurance requirements (i.e., engineering requirements should not be based on insurance requirements).

Following are the two priority research and development topics that fall under this performance-based design for coastal inundation category.

R&D Priority: Performance-based Design for Coastal Inundation

Topic No. 29: Performance levels and acceptance criteria for buildings subject to coastal inundation

Description: There is a lack of tools and guidance available to design and evaluate buildings to withstand coastal inundation on a performance basis beyond what is provided in the NFIP language. A critical first step towards enabling performance- based design is to define expected levels of performance and acceptance criteria for different hazard levels.
Estimated Cost: $500,000 to $1,000,000
Estimated Time: 2-5 years
Measurement Science Challenges and Potential Solutions

Challenges

- Slow pace of change to traditional design methods
- Need cost/benefit tools to provide federal funding support for performance-based design for flood damage reduction

Potential Solutions

✓ Explore solutions used by the Netherlands, Germany and other European countries

Stakeholders and Roles:

Government	Provide technical support and flood insurance data; modify grant programs to support performance-based design solutions
Universities/Research Organizations	Conduct research; develop performance-based design metrics that link to loss reduction
Industry	Develop guidelines that define performance levels and metrics and their application
Standards Organizations	Convert guidelines into standards of practice

Impacts

- Cost-effective performance-based design procedures that, when implemented, result in substantial improvement in building performance during coastal inundation events
- The adoption of performance-based seismic design procedures in codes and standards would provide the potential for reducing coastal inundation hazard losses

R&D Priority: Performance-based Design for Coastal Inundation

Topic No. 30: Sound engineering basis for flood resistant design that is not based on flood insurance requirements

Description: Unlike any other hazard, existing flood guidance and engineering standards are closely tied to NFIP regulations, insurance policy, and flood insurance rate maps. The entire spectrum of flood hazard determination, flood loads, effects and resistance and performance specifications need to be revised from the ground up using engineering principles.
Estimated Cost: > $500,000 to $1,000,000
Estimated Time: > 8 years (Topics 21-29, which encompass flood load determination through building response to flooding, should be completed before this topic can be completed; however, the effort spent in the other areas reduces the cost of accomplishing this topic)
Measurement Science Challenges and Potential Solutions

Challenges

- Must start with a clean slate and no preconceived notions
- Getting flood engineering community to accept a different approach

Potential Solutions

✓ Develop flood design approaches for building codes that are based on engineering methods and get them slowly accepted into the codes or engineering standards

Stakeholders and Roles:	
Government	Provide technical support and modeled flood hazard data
Universities/Research Organizations	Conduct research; develop engineering methods that lead to loss reduction
Industry	Develop guidelines for engineering methods, based on new research, that are not related to flood insurance requirements
Standards Organizations	Convert guidelines into standards of practice

Impacts

- Engineering basis for coastal inundation would be more-risk-consistent with the other hazards
- Achievement of performance-based metrics for coastal inundation could be more closely tied to engineering than happenstance

SECTION 5. PRIORITIZATION AND BENEFITS OF RECOMMENDED R&D ACTIVITIES

This section provides information on the priorities assigned to each research and development topic by workshop participants, as well as a summary overview of the costs and time required to conduct work under each R&D topic. The section concludes with a discussion of the benefits in implementing the recommended R&D activities.

5.1. Prioritization of R&D Topics by Workshop Participants

Table 5-1 summarizes the research and development topics recommended by the June 2012 workshop participants and indicates the priorities assigned to those topics. A combination of methods was used to assign priorities. As topics were developed by breakout groups during the June workshop, each topic was initially assigned a relative importance of high (H), medium (M), or low (L). At the closing plenary session, each workshop participant was asked to identify his/her "top 3" topics out of the 30 recommended R&D topics. None of these 30 had an initial importance rating of low. Participants at the October workshop were briefed on each of these topics, including relative estimated cost and time to complete. They were then asked to vote on which topics they considered to be the most important, irrespective of cost and time required, or the best value, including consideration of importance, cost, and time. The final priorities of highest, high, and medium were assigned using information from both workshops. In addition to the initial ratings of high or medium, topics receiving two or more of the priority designations "top 3", most important (I), or best value (V), were assigned the highest priority. The highest priority topics were generally found to be topics that have been identified frequently in the research needs literature (see Appendix A). Final priorities for each of the 30 topics are shown in the far right-hand column of Table 5-1.

Table 5.1. Prioritization of Recommended R&D Topics

No.	R&D Topic	Subject Area and Program Element*	June Workshop Priorities		October Workshop Priorities		Final Priority
			Initial Importance Category	Top 3	Most Important	Best Value	
Wind Hazard							
1	Hazard maps for high intensity non-synoptic events, including tornadoes	1-2	H	T	I	V	Highest
2	Effects of climate change on extreme wind event characteristics	1-1	M				Medium
3	Quantification of wind-borne debris hazards	1-2	H	T			High
4	Hazard modification by terrain, topography, and nearby buildings	1-1	M				Medium
5	Joint description for hurricane wind, storm surge, and wave hazards	2-5	H				High
Wind Loads and Effects							
6	Pressure coefficients for wind load determination	2-1	H			V	High
7	Computational fluid dynamics tools for practitioners	2-3	H				High
8	Rational approach for wind load analysis	1-1	H	T			High
9	Modeling and simulation for determination of wind loads due to non-synoptic winds, including tornadoes	1-1	H		I		High

No.	R&D Topic	Subject Area and Program Element*	June Workshop Priorities		October Workshop Priorities		Final Priority
			Initial Importance Category	Top 3	Most Important	Best Value	
Wind Resistance							
10	Test methods for resistance to cyclic loading	2-3	M				Medium
11	Field measurement techniques and test methods to evaluate condition of existing construction	2-3	H	T			High
12	Explicit factors of safety for materials and assemblies	1-2	M				Medium
13	Methods or models to predict loss of strength due to aging and deterioration	2-3	H			V	High
14	Guidelines for wind vulnerability assessment of existing buildings.	2-4	H	T	I		Highest
Performance-based Design-Wind							
15	Performance levels and acceptance criteria for wind hazards	1-2	M	T	I		Highest
16	Analysis procedures for nonlinear system behavior used in performance-based design	1-2	H				High
17	Cyber-based tools to support performance based design -	2-3	H	T			High
18	Measurement of windstorm resilience and benefits of performance-based design	2-5	H	T		V	Highest

Table 5.1. (Continued)

No.	R&D Topic	Subject Area and Program Element*	June Workshop Priorities		October Workshop Priorities		Final Priority
			Initial Importance Category	Top 3	Most Important	Best Value	
Coastal Inundation Hazard							
19	Effects of sea level rise and future shoreline changes on coastal inundation hazards	1-1	M				Medium
20	Quantification of flood- borne debris hazards	1-2	H	T			High
21	Effects of over-land flow on waves and other coastal inundation hazards	2-3	H		I		High
22	Characterization of the coastal inundation hazard for low probability events	2-5	H			V	High
Coastal Inundation Load and Effects							
23	Field data collection of flood loads	2-3	H	T	I	V	Highest
24	Load combinations for simultaneous flood hazards	2-3	H				High
25	Validation of flood forces on buildings	2-3	M				Medium
Resistance to Coastal Inundation							
26	Flood damage functions and fragility curves	2-3	H		I		High
27	Effects of erosion and scour on deep foundations	1-2	H			V**	High

No.	R&D Topic	Subject Area and Program Element*	June Workshop Priorities		October Workshop Priorities		Final Priority
			Initial Importance Category	Top 3	Most Important	Best Value	
28	Test methods for materials and systems subject to flooding	2-4	M			V**	Medium
Performance-based Design-Coastal Inundation							
29	Performance levels and acceptance criteria for buildings subject to coastal inundation	1-2	H		I	V	Highest
30	Sound engineering basis for flood resistant design that is not based on flood insurance requirements	2-3	H				High

*Values in cells are provided in x-x format, with the first number indicating either subject area 1 or 2 and the second number indicating program elements 1 through 5, as described in Figure A-1 (Appendix A).

**tie vote for best value topic in Resistance to Coastal Inundation

H, M: High and medium importance, respectively, as assigned by June 2012 workshop participants

I: Most important topic in topic category, irrespective of cost and time required

T: One of ten topics receiving the highest number of "Top 3" votes by workshop participants.

V: Best value in topic category, considering importance, cost, and time

5.2. Summary Overview of Recommended R&D Topic/Activity Costs and Time Requirements

Table 5-2 provides a summary of costs of the recommended activities for each R&D topic, in priority order, along with the estimated costs and time requirements for each topic. The total cost to complete all 30 research and development activities is estimated to be on the order of $35-to-60 million. Assuming annual budgets on the order of $6 million, the complete list of activities would take 6-to-10 years to complete.

5.3. Proposed Program Budget and Schedule for the First Eight Years

The proposed budget (in constant dollars) and schedule for the R&D program for the first eight years of its operation are summarized in Table 5-3. The total budget for each of the 30 R&D activities is based on the "estimated cost" for each topic/activity shown in Table 5-2. In most instances, the proposed budget is taken as the mid-range cost in Table 5-2; in other instances, judgments were made to keep the annual cost for all R&D efforts constant at $6.0 million per year. The costs are spread out over the estimated time the R&D activity is expected to require. The costs are summed for both the wind R&D topics/activities and the coastal inundation R&D topics/ activities; total costs are also shown for the overall R&D program (all topics/ activities).

The overall R&D program starts at a level of $4.0 million per year and grows to $6.0 million per year over the eight year projected time frame. The spending plan through the first eight years illustrates a rapid increase in spending over the first few years.

Table 5.2. Cost and Time Requirements for Recommended R&D Activities

No.	R&D Topic	Priority	Estimated Cost	Estimated Project Duration
	Wind Hazard			
1	Hazard maps for high intensity non-synoptic events, including tornadoes	Highest	$1,000K-3,000K	5-8 years
3	Quantification of wind-borne debris hazards	High	$500K-1,000K	2-5 years
5	Joint description for hurricane wind, storm surge, and wave hazards	High	$1,000K-3,000K	2-5 years
4	Hazard modification by terrain, topography, and nearby buildings	Medium	$500K-1,000K	2-5 years
2	Effects of climate change on extreme wind event characteristics	Medium	$1,000K-3,000K	2-5 years
	Wind Loads and Effects			
6	Pressure coefficients for wind load determination	High	>$3,000K	> 8 years

No.	R&D Topic	Priority	Estimated Cost	Estimated Project Duration
7	Computational fluid dynamics (CFD) tools for practitioners	High	>$3,000K	> 8 years
8	Rational approach for wind load analysis	High	$500K-1,000K	2-5 years
9	Modeling and simulation for determination of wind loads due to non-synoptic winds, including tornadoes	High	>$3,000K	> 8 years
	Wind Resistance			
14	Guidelines for wind vulnerability assessment of existing buildings.	Highest	$500K-1,000K	2-5 years
11	Field measurement techniques and test methods to evaluate condition of existing construction	High	>$3,000K	> 8 years
13	Methods or models to predict loss of strength due to aging and deterioration	High	$500K-1,000K	2-5 years
12	Explicit factors of safety for materials and assemblies	Medium	$500K-1,000K	2-5 years
10	Test methods for resistance to cyclic loading	Medium	$1,000K-3,000K	> 8 years
	Performance-based Design-Wind			
15	Performance levels and acceptance criteria for wind hazards	Highest	$1,000K-3,000K	> 8 years
18	Measurement of windstorm resilience and benefits of performance-based design	Highest	$500K-1,000K	5-8 years
16	Analysis procedures for nonlinear system behavior used in performance-based design	High	$500K-1,000K	5-8 years
17	Cyber-based tools to support performance- based design	High	$1,000K-3,000K	5-8 years
	Coastal Inundation Hazard			
20	Quantification of flood-borne debris hazards	High	$1,000K-3,000K	5-8 years
21	Effects of over-land flow on waves and other coastal inundation hazards	High	$1,000K-3,000K	5-8 years
22	Characterization of the coastal inundation hazard for low probability events	High	$1,000K-3,000K	5-8 years
19	Effects of sea level rise and future shoreline changes on coastal inundation hazards	Medium	$500K-1,000K	5-8 years

Table 5.2. (Continued)

No.	R&D Topic	Priority	Estimated Cost	Estimated Project Duration
Coastal Inundation Load and Effects				
23	Field data collection of flood loads	Highest	>$3,000K	5-8 years
No.	**R&D Topic**	**Priority**	**Estimated Cost**	**Estimated Project Duration**
24	Load combinations for simultaneous flood hazards	High	$500K-1,000K	2-5 years
25	Validation of flood forces on buildings	Medium	$1,000K-3,000K	5-8 years
Resistance to Coastal Inundation				
26	Flood damage functions and fragility curves	High	$500K-1,000K	2-5 years
27	Effects of erosion and scour on deep foundations	High	$500K-1,000K	2-5 years
28	Test methods for materials and systems subject to flooding	Medium	$1,000K-3,000K	5-8 years
Performance-based Design-Coastal Inundation				
29	Performance levels and acceptance criteria for buildings subject to coastal inundation	Highest	$500K-1,000K	2-5 years
30	Sound engineering basis for flood resistant design that is not based on flood insurance requirements	High	$500K-1,000K	> 8 years

Legend: Highest priority R&D topic; Medium priority R&D topic; High priority R&D topic

While there can be large variations in budgets from year to year, based on agency priorities and authorized funding levels, the proposed spending plan provides a possible framework for enacting a Windstorm and Coastal Inundation Impacts Reduction R&D Program.

The scheduling plan (Table 5-3) reflects the proposed priorities for the overall R&D program. Since some of the 30 research and development activities must be done before others, this approximate sequencing also provides a framework identifying which R&D activities should be done or completed first and which should or must follow. The highest priority R&D topics/activities are scheduled to be conducted first, unless their completion is dependent upon the completion of other tasks. Linked tasks are connected with arrows. Only the first eight years are mapped.

Table 5.3. Proposed Windstorm and Coastal Inundation Impacts Reduction R&D Program Schedule and Costs (In thousands of dollars)

	Wind R&D Topics/Activities	Year 1	Year 2	Year 3	Year 4	Year 5	Year 6	Year 7	Year 8	Total
1	Hazard maps for high intensity non-synoptic events, including tornadoes	500	500	500	100	100	100	100	100	2000
2	Effects of climate change on extreme wind event characteristics		550	550	400	300	200			2000
3	Quantification of wind-borne debris hazards				150	150	150	150	150	750
4	Hazard modification by terrain, topography, and nearby buildings				150	150	300	200	100	900
5	Joint description for hurricane wind, storm surge, and wave hazards				400	400	400	400	600	2200
6	Pressure coefficients for wind load determination	500	500	500	500	500	500			3000
7	Computational fluid dynamics tools for practitioners	300	300	300	300	300	300	500	700	3000
8	Rational approach for wind load analysis							200	550	750
9	Modeling and simulation for determination of wind loads due to non-synoptic winds, including tornadoes	500	500	500	300	300	300	300	300	3000
10	Test methods for resistance to cyclic loading			200	250	250	350	500	450	2000
11	Field measurement techniques and test methods to evaluate condition of existing construction		300	300	300	300	300	400	500	2400
12	Explicit factors of safety for materials and assemblies					200	200	200	150	750
13	Methods or models to predict loss of strength due to aging and deterioration	150	150	150	150	150				750
14	Guidelines wind vulnerability assessment of existing buildings.	250	250	250						750
15	Define performance levels and acceptance criteria for wind hazards	250	250	250	250	250	250	250	250	2000
16	Analysis procedures for nonlinear system behavior used in performance-based design		100	100	100	100	100	100	150	750
17	Cyber-based tools to support performance-based design		200	200	300	300	300	300	400	2000
18	Measurement of windstorm resilience and benefits of performance-based design	100	100	100	150	150	150			750
	Wind R&D Activities Total	**2550**	**3700**	**3900**	**3800**	**3900**	**3900**	**3600**	**4400**	**29750**

From Topic 22 (Characterization of the coastal inundation hazard for low probability events)

Table 5.3. (Continued)

	Coastal Inundation R&D Topics/Activities	Year 1	Year 2	Year 3	Year 4	Year 5	Year 6	Year 7	Year 8	Total
19	Effects of sea level rise and future shoreline changes on coastal inundation hazards				100	100	100	100	350	750
20	Quantification of flood-borne debris hazards					500	500	750	750	2500
21	Effects of over-land flow on waves and other coastal inundation hazards	400	400	400	500	300				2000
22	Characterization of the coastal inundation hazard for low probability events	400	400	450	400	350				2000
23	Field data collection of flood loads	700	750	750	750	50				3000
24	Load combinations for simultaneous flood hazards						250	250		500
25	Validation of flood forces on buildings					500	500	500		1500
26	Flood damage functions and fragility curves						250	250		500
27	Effects of erosion and scour on deep foundations						200	350		550
28	Test methods for materials and systems subject to flooding		300	300	350	300	300			1550
29	Define performance levels and acceptance criteria for buildings subject to coastal inundation	200	200	200	100					700
30	Develop sound engineering basis for flood resistant design separate from flood insurance requirements							200	500	700
	Coastal Inundation R&D Activities Total	**1700**	**2050**	**2100**	**2200**	**2100**	**2100**	**2400**	**1600**	**16250**
	Wind R&D Activities Total (prior page)	**2550**	**3700**	**3900**	**3800**	**3900**	**3900**	**3600**	**4400**	**29750**
	R&D Program Total (All Activities)	**4250**	**5750**	**6000**	**6000**	**6000**	**6000**	**6000**	**6000**	**46000**

To Topic 5 (Joint description for hurricane wind, storm surge, and wave hazards)

Legend: Highest priority R&D topic; High priority R&D topic; Medium priority R&D topic; → Activity link pointing to sequential activity

It is recommended that the Windstorm and Coastal Inundation Impacts Reduction R&D Program be implemented at the earliest possible moment. Start-up should be rapid, since there has already been a significant amount of effort (in this Roadmap development effort) in defining what is to be done and how it can be accomplished.

5.4. Benefits of Implementing R&D Activities for Windstorm and Coastal Inundation Loss Reduction

The benefits of the recommended R&D program include:

- Better, more technically sound wind- and coastal inundation-resistant engineering design practices;
- More reliable structures that perform better in high winds and flooding;
- Fewer lives lost in a destructive windstorm (especially tornadoes) or coastal inundation event;
- Lower initial and retrofit costs related to wind- and/or coastal inundation- resistant construction;
- Mitigation of the consequences of windstorms and coastal events; and
- Better written, more easily understood codes and standards.

For the nation as a whole, implementation of the proposed R&D program will yield the following major benefits:

1. Reduction in the traumatic life loss, injury, damage, and economic impacts when windstorms and coastal inundation events occur.
2. More rapid recovery and restoration of the physical community and economic activity following a significant windstorm and/or coastal inundation event.
3. Reduced investment required to achieve risk-consistent design and construction of buildings subject to wind and coastal inundation events

The accrual of benefits to the design practice will begin almost immediately once the program has begun. Researchers and practitioners in both the wind and coastal engineering communities have requested research help for numerous important technical issues for many years. Implementation of the proposed R&D program will immediately signal an important positive change to the technical community that should spark new enthusiasm for pursuing worthy research and developmental efforts that will improve current knowledge about both the windstorm and coastal inundation hazards and ways to reduce significantly the impacts of these hazards.

Appendix A. Research Needs Literature

A.1. Literature Review of Existing Resources and Past Workshops on Identifying Research Needs

The review of the research needs literature included identification of the key documents from the past 20 years and extracting research recommendations relevant to the measurement science roadmap. Following is a list of documents reviewed. The full citation for each document is provided in the list of References.

[1] *Hurricane Warning: The Critical Need for a National Hurricane Research Initiative*, National Science Board of the National Science Foundation, January 2007.

[2] *Windstorm Impact Reduction Implementation Plan*, National Science and Technology Council, April 2006.

[3] *Wind Engineering Research and Outreach Plan*, American Association for Wind Engineering and the American Society of Civil Engineers (ASCE), May 2004.

[4] *New Opportunities to Reduce Wind Hazard Losses and Improve Quality of Life in the USA*, American Association for Wind Engineering, August 1997.

[5] *Proceedings of International Workshop on Wind Engineering Research and Practice: Current State of the Art and Future Needs/Plans*, Chapel Hill, NC, May 2010.

[6] *NOAA's Storm Surge Roadmap: Transition Research to NWS Operations*, J. Feyen, July 2010.

[7] *Coastal Storm Modeling – System Integration*, US Army Corps of Engineers, T. Massey, T. Wamsley, M. Cialone, 2011;

[8] *Wood Engineering Challenges in the New Millennium – Critical Research Needs*, Proceedings of the Workshop Offered in Conjunction with the Structural Engineering Institute (SEI) of ASCE Structures Congress 2008, V. K. A. Gopu, April 2008.

[9] *Performance Based Design Extreme Wind Loads on a Tall Building, The Structural Design of Tall Buildings*, A. Jain, M. Srinivasan, G. C. Hart, August 2000.

[10] *Toward the Probabilistic Simulation of Storm Surge and Inundation in a Limited-Resource Environment*, Davis et al, American Meteorological Society, 2010.

[11] *Grand Challenges for Disaster Reduction*, National Science and Technology Council, Subcommittee on Disaster Reduction, 2005.

[12] *Proceedings, Workshop on Research Needs in Wind Engineering*, NISTIR 5597, National Institute of Standards and Technology, 1995.

[13] *Windstorm Mitigation Initiative, Retrofit Workshop Report*, Texas Tech University, December 1999.

[14] *Grand Challenges for Disaster Reduction*, National Science and Technology Council, Subcommittee on Disaster Reduction, 2008.

[15] *Coastal Hazards Colloquium*, Department of Homeland Security, Science and Technology Directorate, University of North Carolina, December 2008.

[16] *Wind Engineering Research Needs*, P. Irwin, ASCE Technical Council on Wind Engineering, 2011.

[17] *Disaster Resilience: A National Imperative*, National Academies Press, 2012.

[18] *Developing Guidelines and Standards for Disaster Resilience of the Built Environment: A Research Needs Assessment*, NIST Technical Note 1795, March 2013.

[19] *Wind and the Built Environment, U.S. Needs in Wind Engineering and Hazard Mitigation*, National Research Council, 1993. and

[20] *ATC-57, The Missing Piece: Improving Seismic Design & Construction Practices*, Applied Technology Council, 2003.

A.2. Categorization of Reviewed Documents by Initial Set of Program Subject Areas and Elements

The recommendations from the literature were organized under the initially identified two subjects areas and five program elements (as adapted from the ATC-57 Report) that were included in the work plan for this roadmap development effort (Figure A-1). Under each program element, a brief summary is provided, followed by a list of specific recommendations, where the number in [ ] before each entry in the list refers to the number assigned to each reference in Section A.1.

A.2.1. Subject Area 1: Systematic Support of the Windstorm and Coastal Inundation Code Development Process

Program Element 1: Provide technical support for windstorm and coastal inundation engineering practice and code development process.

The literature focused on modeling of hazards for the purposes of determining loading requirements and the simulated effects of the hazards on the structures. Given the variety of building shapes, and the lack of more precise knowledge about the specific impacts of wind loading on various structural and non-structural elements, the literature also suggests that simulation and modeling techniques for wind must be created in order for practitioners to develop cost-effective wind designs. These techniques should include numerical modeling, visualization, database-assisted design, and wind simulations.

The literature that focused on codes and standards suggested that improvements were needed in the number of building shapes for which wind pressure coefficients exist and that risk and vulnerability tools were needed for assessing the probable performance of critical facilities and infrastructure.

[3] Develop simulation techniques for wind modeling;
[3] Develop database and knowledge-based models for wind loading on structures;
[3] Develop simulation techniques for modeling of wind loading on structures;
[3] Develop techniques for modeling effects of wind hazards;
[3] Improve characterization of properties of severe winds;
[3] Develop instrumentation and data transfer/processing infrastructure for wind data acquisition;
[3] Develop computational methods for simulating wind loading;
[3] Develop databases and visualization tools for simulation of wind effects;
[4] Modeling of wind-structure interaction, including effects of integral wind loads on structural systems, components and cladding, effectiveness of retrofitting schemes, effects of structural fatigue and impact by wind-generated missiles, design of cost effective tornado shelters and shelters for hurricane zones to minimize evacuation;
[4] Application of effective numerical schemes using computational fluid dynamics to determine the wind environment and wind loading on and response of buildings, structures, transportation systems and other critical components of civil engineering infrastructure, and to mitigate these effects;
[5] Study the possibilities of using wind database-assisted design techniques; [5] Study the possibilities of using wind database-assisted design techniques; [5] Need new gust loading factors, including 3-D gust loading factors;
[5] Need wind design methodology for photovoltaic systems;
[5] Investigate shape coefficients for more roof shapes than currently exist in ASCE 7, *Minimum Design Loads for Buildings and Other Structures*;
[13] Develop tools to simulate aging of components including corrosion; [13] Develop methods for assessing load path continuity;

- ***Systematic Support of the Windstorm and Coastal Inundation Code Development Process***

  Program Element 1: Provide technical support for windstorm and coastal inundation engineering practice and code development process.

  Program Element 2: Develop the technical basis for performance-based windstorm and coastal inundation engineering by supporting problem-focused, user-directed research and development.

- ***Improve Windstorm and Coastal Inundation Design and Community Resilience***

  Program Element 3: Support the development of technical resources (e.g., guidelines, manuals, software tools) to improve windstorm and coastal inundation engineering practice.

  Program Element 4: Make evaluated technology available to practicing professionals in the windstorm and coastal inundation design and construction communities.

  Program Element 5: Develop the technical basis for windstorm and coastal inundation engineering to support community resilience within an all-hazards framework.

Figure A.1. Program subject areas and elements as defined in the work plan for this Roadmap development effort.

[14] Provide the technical basis for revised codes and standards for critical infrastructure and essential facilities by using risk and vulnerability assessment tools;
[16] Increase the number of building shapes available in the codes and standards;
[19] Develop or improve analytical, numerical and experimental methods for determining wind loading; and
[19] Develop comprehensive wind speed database.

Program Element 2: Develop the technical basis for performance-based windstorm and coastal inundation engineering by supporting problem-focused, user-directed research and development.

The literature focused on defining performance measures for structural frames and the building envelope for the wind hazard. Performance levels should be defined for varying levels of hazard intensity. The design criteria should reflect system reliability issues, not component reliability.

Several specific problem-focused studies were mentioned in the literature, including the need to better understand both wind- and flood-borne debris, serviceability thresholds, and the changes that occur in both wind fields and coastal inundation levels at the land-sea interface.

[1] Study the interaction of hurricanes with engineered structures;
[2] Evaluate the response of the built environment and critical infrastructure to wind events by investigating aerodynamic response, load path, ultimate capacity and the performance of the building envelope;
[2] Assess the impact of wind and windborne debris or wind and water/ice/snow; [3] Improve understanding of wind-borne debris;

[3] Define performance measures;
[4] Conduct research on debris impact potential in windstorm and develop impact resistant building components;
[5] Research the transition in turbulence from Exposure D to Exposure B at landfall; [5] Determine serviceability thresholds for wind motion in tall buildings;
[5] Conduct research on wind-borne debris;
[8] Develop and implement performance-based approaches for the design of wood structures;
[11] Develop reliable methods to design structures to meet specific performance levels under increasing levels of hazard intensity;
[12] Establish design criteria that reflects system reliability as opposed to component reliability; and
[13] Determine characteristics of wind-borne missiles, including better definition of the geographical boundaries where missile impacts should be taken into account in design, and determination of the correlation between missile size, weight and impact speed with wind speed.

A.2.2. Subject Area 2: Improve Windstorm and Coastal Inundation Design and Community Resiliency

Program Element 3: Support the development of technical resources (e.g., guidelines, manuals, software tools) to improve windstorm and coastal inundation engineering practice.

The literature focused on the development of guidance for wind and coastal inundation issues that could be modeled after the guidance already developed for other hazards, such as earthquakes. Specifically, the literature mentioned loss assessment methodologies, damage and fragility models, response of structural systems and probabilistic methods as examples of issues that have been studied and written about for earthquake practitioners that could be carried forward and studied for wind and flood practitioners.

The literature also suggested that some tools specific to coastal inundation should be made available to practitioners, such as a storm surge model that could be used to develop surge scenarios or could be used specifically for design and an improved method for determining water levels or depths at any point in the flooded area.

[2] Explore the near-ground and channeling/shielding effects of winds on buildings through testing and instrumentation
[2] Measure the response of bridges and other highway structures to wind events, including stability, serviceability and functionality leading up to and through extreme events
[3] Improve knowledge on behavior of structural and non-structural components;
[3] Improve understanding of response of structural systems;
[3] Improve damage and fragility models for structures;
[3] Improve indirect loss estimation models for wind hazards;
[3] Develop real-time loss estimation tools;
[4] Study internal load paths and the performance of structural systems, and effectiveness of connections between structural components;
[6] Develop the Advanced Circulation (ADCIRC) model for operational use;

[6] Improve determination of storm water levels;
[6] Develop techniques to show flood depth above ground;
[11] Improve loss assessment methodologies and decision support tools to include multiple hazards, based on existing tools developed for earthquake engineering;
[13] Improve probabilistic treatment of the variability in loads and resistance;
[14] Develop probabilistic inundation hazards prediction and methods to effectively quantify and communicate risk; and
[19] Develop effective technology transfer techniques to apply research results to design of wind-resistant structures and components.

Program Element 4: Make evaluated technology available to practicing professionals in the windstorm and coastal inundation design and construction communities.

The literature focused primarily on developing technologies since there have been few resources available for pursuing technology to the point where it could be made available and useful to practitioners. Some of the topics would appear to need some significant study before tools can be developed for use by practitioners. However, once available, the information could be of significant use to the practice with the potential to reduce design and construction costs of buildings built in wind and coastal inundation areas and to reduce losses caused by these events.

The topics discussed in the literature include hazard identification issues, such as improved mapping of the wind hazard; loading issues, such as the effects of various types of windstorms on structural loading; and the very dynamic transition that occurs in both wind and water loads at the interface between the water and the land. Also included were topics dealing with how to better determine the appropriate retrofitting requirements for existing buildings.

[1] Develop improved techniques for in situ observations;
[1] Develop a better understanding of air-sea interaction;
[2] Develop new technologies and ground, airborne and satellite based storm observation systems to improve knowledge and understanding of windstorms and the wind variability within those storms;
[2] Examine the interaction between wind and storm surge to determine the impact on building foundations and critical infrastructure;
[3] Improve mapping of wind hazard;
[3] Develop tools for real-time monitoring of wind hazards;
[3] Continue to develop remote-sensing technologies for damage assessment;
[3] Develop retrofit techniques for new and existing materials, components systems and subsystems;
[3] Demonstrate application of new tools in post-event setting;
[4] Collect wind speed data using robust instrumentation and state-of-the-art technology to map the detailed structure of the wind, topographic effects, and long-term climate effects;
[4] Develop effective techniques for collection, rapid archiving, and dissemination of data acquired during post-disaster investigations;
[4] Develop cost-effective retrofit techniques to enhance wind resistance of existing structures;

[4] Map wind climate in urban areas;
[4] Simulate hurricanes and their wind fields, and other extreme wind effects for statistical analysis of wind, wind loads, and wind-induced response of structures and their components;
[5] Study the variations of results of wind tunnel studies;
[5] Develop/improve damping technology for tall buildings;
[5] Study the effects of exposure and terrain on wind pressures at buildings;
[5] Study the effects of building size and shape on attenuating wind pressure;
[5] Study the flow around buildings from tornado vortex to determine pressure requirements for design;
[7] Improve storm surge modeling by coupling five other coastal models to the U.S. Army Corps of Engineers' Engineering Research and Development Center's Coastal Storm Modeling System (CSTORM-MS);
[13] Develop an understanding of long-term performance of glues, adhesives and coatings;
[14] Examine the interaction between wind, storm surge and shallow water waves to determine the impact on building foundations, critical infrastructure, and vegetation;
[14] Evaluate capabilities of radar for swell/wave and deformation measurements;
[14] Develop coastal wind/wave climate maps and shoreline coastal process models to better understand and predict aspects of coastal erosion and inundation;
[14] Determine climate scale factors that relate to sea level variability and rise;
[15] Improve quantitative predictions of coastal erosion;
[15] Improve understanding of the physics of tropical cyclones or coastal ocean conditions at the land-water interface;
[16] Study the effects on structures of wind loading caused by various wind event types;
[16] Study the effects of building surroundings on structural reliability and risk of damage caused by wind loading;
[16] Establish a program of real-time wind monitoring of buildings;
[19] Better define wind loads in codified form for use with small buildings;
[19] Research windflow around buildings to improve empirical prediction of cladding pressures;
[19] Improve knowledge of dynamic loading effects on flexible structures; and
[19] Research cyclic loading to improve understanding of fatigue problems.

Program Element 5: Develop the technical basis for windstorm and coastal inundation engineering to support community resilience within an all-hazards framework.

The literature had a significant number of references to improvements that are needed in assessing community resiliency including determining the benefits to the community for community-wide loss reduction measures. There are several references to the need for technologies to aid in the rapid repair and restoration of critical infrastructure and for material development to improve long-term building performance and reduce construction cost.

The literature also articulated the need to develop hazard identification tools for multiple hazards across the broad spectrum of what could constitute a hazard for any particular community.

[1] Develop better capabilities to predict hurricane intensification and size, and thereby reduce the uncertainty associated with where and when hurricanes will make landfall;
[1] Study the economic and social impacts of hurricanes and develop new mitigation measures;
[1] Assess and improve the resilience of the built environment;
[2] Develop and implement technologies for rapid repair and restoration of critical infrastructure and critical services;
[2] Assess individual and community capability to respond to wind events, including vulnerability analyses, risk perception, risk communication, risk management, communication of wind warnings and public response, evacuation capability, and public knowledge of appropriate protective actions for wind events, especially among vulnerable populations;
[3] Develop methodologies and assess effectiveness in enhancing community resilience to the wind hazard;
[4] Develop and apply reliable techniques for cost-benefit analysis of wind hazard mitigation measures and other means for socio-economic evaluations;
[5] Study the interface between wind engineering and green building design;
[8] Develop next generation wood systems, including materials, connections, assemblies and products;
[8] Develop advanced design and construction methodologies using combined materials (hybrid construction);
[11] Identify mitigation strategies and technologies that can provide simultaneous protection against more than one hazard for a single cost; and
[11] Develop technologies to prevent cascading failures of complex lifeline systems.

Table A.1. Research Priorities from Literature Search

Research Topic	Number of Times Mentioned	Program Element Mapping*
Wind Hazard		
Develop modeling or simulation techniques for wind hazard	7	1-1
Develop better understanding of wind-borne debris	4	1-2
Improve wind mapping	3	2-4
Wind Loads		
Develop computational models or use databases for wind loading	12	1-1
Develop wind pressure coefficients for an increased number of building shapes and systems	2	1-1
Study terrain, exposure factors in developing wind pressures	1	2-4
Collect real-time wind pressure measurements to calibrate models	7	2-4
Wind Resistance		
Develop capability to assess load path continuity	4	1-1
		2-3

Research Topic	Number of Times Mentioned	Program Element Mapping*
Develop tools to simulate aging	1	1-1
Improve loss estimation tools, including fragility models	5	2-3
Develop retrofit techniques	2	2-4
Performance-based Design Wind		
Define performance levels and acceptance criteria for the wind hazard	4	1-2
Coastal Inundation Hazard		
Improve techniques for finding flood depths, storm surge inundation extent, and probabilistic flood depths	5	2-3
Coastal Inundation Loads		
Improve knowledge of erosion, climate changes, and wave impacts	5	2-4
Coastal Inundation Resistance Performance-based Design & Coastal Inundation Resiliency and Sustainability		
Improve performance of infrastructure	3	2-5
Research new materials and assemblies	4	2-5
Develop cost-effectiveness and reliability assessment methods	5	2-5

*Values in cells are provided in x-x format, with the first number indicating either subject area 1 or 2 and the second number indicating program elements 1 through 5 (See Figure A-1 for program subject area and element definitions).

A.3. Research Priorities Identified from the Literature

The literature search covered approximately 20 years of reports and studies on the need for wind and coastal inundation research and each of these studies in some way identified specific research topics. The literature used for this analysis is only that which represented the views of many reviewers, workshop participants, or authors of the document. The number of times the various research and development recommendations identified above were cited in the literature is summarized in Table A-1 as a measure of their importance and/or longevity of need. The topics are organized by hazard type (wind or coastal inundation) and then by hazard subject area (hazard identification, loads, resistance, and performance-based design.

Appendix B. June 2012 Reston, Virginia Workshop

B.1. Workshop Agenda

Wednesday, June 13, 2012

8:00 - 8:30 am Registration, Continental Breakfast

8:30 - 8:40 am	Welcome, Housekeeping, Plan for the 2 Days – Bill Coulbourne (ATC)
8:40 - 9:30 am	Overview of NIST and Engineering Laboratory – and Charge to Workshop Participants - Shyam Sunder (NIST)
9:30 - 9:45 am	NSF's Role in Research - Kishor Mehta (NSF)
9:45 - 10:00 am	Introduction of the Research Focus Areas – Marc Levitan (NIST)
10:00 - 10:15 am	Break
10:15 - 10:45 am	1st Technical Plenary – Disaster Resilience of Building, Infrastructure, and Communities– Steve Cauffman (NIST)
10:45 - 10:50 am	Instructions for Morning Breakout Sessions – Bill Coulbourne (ATC)
10:50 - 11:50 am	Break-Out Sessions - Future Vision of Resilient Communities and How We Get There

For the first breakout sessions, each pre-assigned group will define what they think constitutes a disaster resilient community, how this community would be identified or defined, and the gaps in current knowledge that need to be addressed to enable disaster-resilient communities.

11:50 - 12:20 pm	Resilience Groups Report Out (*Plenary Meeting Room)*
12:20 - 1:00 pm	Lunch
1:00 - 1:45 pm	2nd Technical Plenary – Moving Toward Performance-Based Design for Wind: Larry Griffis (Walter P. Moore) *Chris Letchford to introduce Larry Griffis*
1:45 - 2:15 pm	3rd Technical Plenary – R&D Challenges for Coastal Inundation Impact Reduction: Gary Chock (Chock and Martin) *Chris Jones to introduce Gary Chock*
2:15 - 2:20 pm	Instructions for Afternoon Breakout Sessions – Bill Coulbourne (ATC)
2:20 - 5:05 pm	Break-Out Sessions – by Discipline Groups

Breakouts into the five pre-assigned discipline areas listed below for identification of gaps in current knowledge and research needs. Use literature search document as background for gaps in knowledge and research needs identified by other groups over the last 10-15 years. Groups also discuss possible Grand Challenge needs.

(Moderators for each group shown in italics)

Room A – Wind hazard group *(Chris Letchford)*
Room B – Wind loading group *(Jon Galsworthy)*
Room C – Wind resistance group *(Tom Smith)*
Room D – Flood hazard group *(Chris Jones)*
Room E – Flood loading and resistance group *(Bill Coulbourne)*

3:15 - 3:30 pm	Break (take at leisure)
5:05 - 5:15 pm	Organization for Tomorrow – Bill Coulbourne (ATC) *Plenary Meeting Room*

5:15 pm	Adjourn *Shuttle buses will take participants back to the Sheraton Hotel*

Thursday, June 14, 2012

8:00 - 8:30 am	Continental Breakfast
8:30 - 9:40 am	Discipline Groups Report Out Including Grand Challenge Ideas *Plenary Meeting Room*
9:40 - 9:45 am	Instructions for Morning Breakout Sessions – Bill Coulbourne (ATC)
9:45 - 11:45 am	Break-Out Sessions – Crosscutting by Program Element

Groups will change for Day 2 as the focus will shift. Participants are pre-assigned to one of the five areas listed below; corresponding to the program elements adapted from ATC-57 approach and will be provided summaries of the Resilience and Disciplinary breakout group research needs lists. Each group must identify the Top 10 research needs for their assigned program element. (Moderator for each breakout group shown in italics)

Room A – *Provide technical support for windstorm and coastal inundation engineering practice and code development process.* What specific technical support is needed by the committees that develop model codes and the standards upon which the model codes depend? (*Chris Letchford)*

Room B – *Develop the technical basis for performance-based windstorm and coastal inundation engineering by supporting problem-focused, user-directed research and development.* What research is required to develop the technical basis for performance-based design for windstorms and coastal inundation? (*Jon Galsworthy)*

Room C – *Support the development of technical resources (e.g., guidelines and manuals) to improve windstorm and coastal inundation engineering practice.* What new technical resources are needed by practitioners? What research and development must be done in order to provide these resources? (*Chris Jones)*

Room D – *Make evaluated technology available to practicing professionals in the windstorm and coastal inundation design and construction communities.* In what ways should evaluated technology be made available to practicing professionals in the windstorm and coastal inundation design and construction communities? On what current topics is significant research reported in the literature but needs to be evaluated and coalesced into a single document designed for use by practitioners. (*Tom Smith)*

Room E – *Develop the technical basis for windstorm and coastal inundation engineering to support community resilience within an all-hazards framework.* Building on the discussion in Day 1 on community resilience, what research is needed to define, measure, and provide the basis for improvements to the resilience of buildings, infrastructure, and communities for windstorms and coastal inundation? (*Horia Hangan)*

10:00 - 10:15 am	Break (take at leisure)
11:45 - 12:15 pm	Lunch
12:15 - 2:00 pm	Report Out and Discuss Top 10 Research Projects per Program Element *Plenary Meeting Room*

2:00 - 2:30 pm	Breakout Session - Fine-Tuning Research Priorities *Return to morning breakout rooms*
2:30 - 2:45 pm	Break
3:00 - 3:55 pm	Discuss and Prioritize Overall Top 15 Research Projects *Plenary Meeting Room*
3:55 - 4:00 pm	Workshop Closing Remarks – Marc Levitan (NIST)
4:00 pm	Adjourn *Shuttle buses will take participants back to the Sheraton Hotel*

B.2. Workshop Presentations

The following is a synopsis of each of the three technical presentations delivered in the June workshop.

- *Disaster Resilience of Building, Infrastructure, and Communities* by Stephen Cauffman (NIST)

This presentation focused on defining resilience especially at the community level. Resilience was described as the capability of a system to maintain acceptable levels of functionality during and after a disruptive event and to recover fully within some specified time period. Issues were identified to help achieve resilient communities. An example was given of what the City of San Francisco has done to develop performance goals for their inventory of buildings and infrastructure. New design codes and standards are needed that deal with performance objectives.

- *Moving Toward Performance-Based Design for Wind* by Larry Griffis (Walter P Moore)

This presentation covered the state of current structural engineering practice for wind and compared it to potential benefits if the practice were modified using performance-based design concepts. The current state of the practice is to design to yield strength instead of extending the load into the elastic response range. The current practice thus uses a static elastic analysis procedure instead of considering the dynamic effects of wind and using a non-linear dynamic analysis procedure using performance objectives to establish the limits of allowable motion. ASCE 41-06 *Seismic Rehabilitation of Existing Buildings* was used as guide for the comparative analysis.

- *R&D Challenges for Coastal Inundation Impact Reduction* by Gary Chock (Martin & Chock)

This presentation began with the general effects of flood inundation and used earthquakes and the effects of tsunamis as the basis for the R&D challenges since storm surge impacts are similar to tsunamis. The R&D needs were defined as: coastal storm flood depths and velocity maps with mean recurrence interval consistent with wind standards; tsunamis inundation design maps; a better understanding of tsunami drawdown flow, which can be very significant; research on scour effects of water flow around buildings; a better understanding of combined structural loadings on building and non-building structures; research on post-elastic structural response of extreme hydrodynamic forces; and structural response metrics for performance-based tsunami design.

B.3. Workshop Participants

Table B.1. June 2012 Workshop Participants

Last Name	First Name	Organization
Adams	Stuart	Louisiana State University
Annane	Bachir	University of Miami
Baskaran	Bas	National Research Council
Bienkiewicz	Bogusz	Colorado State University
Bitsuamlak	Girma	Western University
Bosch	Harold	Federal Highway Administration (HRDI-50)
Bres	Dana	U.S. Dept. of Housing and Urban Development
Caracoglia	Luca	Northeastern University
Cauffman	Stephen	NIST-Engineering Laboratory
Chock	Gary	Martin & Chock, Inc.
Chowdhury	Arindam	Florida International University
Cocke	Elizabeth	U.S. Dept. of Housing and Urban Development
Coulbourne	William	Applied Technology Council
Cox	Daniel	Oregon State University
Dickinson	Tamara	Office of Science and Technology Policy
Dregger	Philip	Technical Roof Services, Inc.
Ehrlich	Gary	National Association of Home Builders
Feyen	Jesse	National Oceanic and Atmospheric Administration
Francis	Mathew	URS Corp.
Friedland	Carol	Louisiana State University
Friis	Donna	CDM Smith
Galsworthy	Jon	Rowan Williams Davies & Irwin, Inc.
Goupil	Jennifer	Structural Engineering Institute of ASCE
Griffis	Lawrence	Walter P Moore
Grosshandler	William	NIST-Engineering Laboratory
Hajj	Muhammad	Virginia Tech
Hall	Tom	Natural Hazards Zurich North American
Hangan	Horia	Western University
Harvey	Doug	Building Officials Association of Florida
Jones	Christopher	Christopher P. Jones & Associates
Kareem	Ahsan	University of Notre Dame
Kennedy	Andrew	University of Notre Dame
Laatsch	Edward	Department of Homeland Security Federal Emergency Management Agency

Table B.1. (Continued)

Last Name	First Name	Organization
Letchford	Chris	Rensselaer Polytechnic Institute
Letvin	Eric	NIST-Engineering Laboratory
Levitan	Marc	NIST-Engineering Laboratory
Lin	Ning	Massachusetts Institute of Technology
Lombardo	Franklin	NIST-Engineering Laboratory
Lynett	Patrick	University of Southern California
Mehta	Kishor	National Science Foundation
Ockerman	Darwin	U.S. Geological Survey
Pauschke	Joy	National Science Foundation
Phan	Long	NIST-Engineering Laboratory
Prevatt	David	University of Florida
Reed	Dorothy	University of Washington
Reinhold	Timothy	Insurance Institute for Business & Home Safety
Resio	Don	University of North Florida
Rojahn	Christopher	Applied Technology Council
Sadek	Fahim	NIST-Engineering Laboratory
Sarkar	Partha	Iowa State University
Schroeder	John	Texas Tech University
Scott	Donald	PCS Structural Solutions
Shackelford	Randall	Simpson Strong-Tie Co.
Shyam-Sunder	Sivaraj	NIST-Engineering Laboratory
Simiu	Emil	NIST-Engineering Laboratory
Smith	Douglas	Texas Tech University
Smith	Thomas	TLSmith Consulting, Inc.
Vickery	Peter	Applied Research Associates
Yeo	DongHun	NIST-Engineering Laboratory

APPENDIX C. OCTOBER 2012 MIAMI, FLORIDA WORKSHOP

C.1. Workshop Description

A second workshop was held in Miami on October 26, 2012, following the conclusion of the ATC/SEI *Advances in Hurricane Engineering Conference*, where many practitioners and academics were already gathered discussing the latest research on hurricanes, storm surge, and broader aspects of wind and coastal engineering. The objective of the October Wind and Coastal Inundation Research Needs Workshop was to refine the results of the workshop held in June 2012. The October workshop was held in two sessions (each 90 minutes in length), with 36 participants in the windstorm session and 15 in the coastal inundation session.

Participants were briefed on the roadmap project and the results of the June workshop. Each research topic in the areas of hazard, loads, resistance, and performance-based design was discussed by the participants. The participants had the opportunity to provide comments on the research topics, input from which is reflected in Section 4.

At the conclusion of each workshop session, participants were provided a "ballot" and asked to select in each of the four topic areas (hazard identification, loads, resistance, and performance-based design) the research item that they considered to be the most important (irrespective of time and cost) and also the topic that they considered to be the best value (including consideration of importance, time and cost). These rankings are indicated in Table 5-1.

C.2. Workshop Participants

Table C.1. October 2012 Wind Workshop Participants

Last Name	First Name	Affiliation
Adams	Stuart	Louisiana State University
Aponte	Luis	University of Puerto Rico at Mayaguez
Baradaranshoraka	Mohammad	Louisiana State University
Boggs	Daryl	CPP, Inc.
Cai	Steve	Louisiana State University
Carter	Russell	Texas Tech University
Charney	Finley	Virginia Tech
Chock	Gary	Martin & Chock, Inc.
Cochran	Leighton	Leighton Cochran Consulting
English	Elizabeth	University of Waterloo
Friedland	Carol	Louisiana State University
Galsworthy	Jon	RWDI
Griffis	Larry	Walter P Moore
Jain	Anurag	Weidlinger Associates, Inc.
Karns	Jesse	MiTek Ind.
Lavelle	Francis	Applied Research Associates
Liu	Fanggian	Clemson University
Mason	Matthew	Risk Frontiers - Macquarie University
Masters	Forrest	University of Florida
Matthews	Elizabeth	Louisiana State University
Mehta	Kishor	National Science Foundation
Miltiades	Ted	Georgia Department of Community Affairs
Nelson	Rawn	MiTek Ind.
Pan	Fang	Louisiana State University
Pei	Bin	Clemson University
Peterka	Jon	CPP, Inc.
Prevatt	David	University of Florida
Rojahn	Christopher	Applied Technology Council
Roueche	David	University of Florida
Ruark	Dean	PGT Industries
Scott	Donald	PCS Structural Solutions

Table C.1. (Continued)

Last Name	First Name	Affiliation
Shafieezadeh	Abdollah	Ohio State University
Smith	Douglas	Texas Tech University
Smith	Mario	None provided
Smith	Tom	TLSmith Consulting
Towson	Bill	Georgia Department of Community Affairs

Table C.2. October 2012 Flood Workshop Participants

Last Name	First Name	Affiliation
Adams	Stuart	Louisiana State University
Aponte	Luis	University of Puerto Rico at Mayaguez
Chock	Gary	Martin & Chock, Inc.
Cochran	Leighton	LCC Ltd.
English	Elizabeth	University of Waterloo
Friedland	Carol	Louisiana State University
Hayden	Jesse	URS Corporation
Matthews	Elizabeth	Louisiana State University
Mehta	Kishor	National Science Foundation
Miltiades	Ted	Georgia Department of Community Affairs
Ogea	Shandy	Louisiana State University
Rojahn	Christopher	Applied Technology Council
Shahieezadeh	Abdollah	Ohio State University
Smith	Douglas	Texas Tech University
Towson	Bill	Georgia Department of Community Affairs

APPENDIX D. WORKSHOP RESEARCH NEEDS MAPPED TO PROGRAM ELEMENTS

Breakout groups at the June 2012 workshop identified a number of research and development needs for each of the initial program elements, as defined in the work plan for this Roadmap development effort. These research and development needs are identified in the subsections that follow. Within each program element, the groups were also asked to identify the three most important topics, which are indicated in italics at the beginning of each list and factored into the selection of the priorities provided in Section 5. The designation of "maps to" indicates how the item corresponds to the numbered R&D topics in Sections 4 and 5, for those topics that were ultimately included.

D.1.Subject Area 1: Systematic Support of the Windstorm and Coastal Inundation Code Development Process

Program Element 1: Provide technical support for windstorm and coastal inundation engineering practice and code development process.

- *Tornado loads (maps to R&D topic 9)*
- *Develop & validate test methods for assemblies (wind, rain, debris, wave) (maps to R&D topic 11)*
- *Update and validate wind/wave/debris loading (maps to R&D topic 20)*
- Characterize wind/coastal flooding hazards (probability distribution) and joint probability distribution (maps to R&D topic 1)
- Impact of climate change on wind/wave hazard (maps to R&D topic 2)
- Hazard modification by terrain/topography/buildings (maps to R&D topic 4)
- Uncertainty analysis of systems (maps to R&D topic 5)
- Aging and repeated loading (inter & intra hazard events), (maps to R&D topic 11)

Program Element 2: Develop the technical basis for performance-based windstorm and coastal inundation engineering by supporting problem-focused, user-directed research and development.

- *Develop the hazard maps for wind and hydrodynamic loading (maps to R&D topic 1)*
- *Define performance levels by hazard (e.g., wind, hydrodynamic loads) (maps to R&D topic 15)*
- *Develop loading & structural analysis procedures and validation (maps to R&D topics 8 and 23)*
- Develop debris hazard maps (wind & water) (maps to R&D topics 3 and 19)
- Define acceptability criteria by performance level (maps to R&D topic 15)
- Study non-linear/post-elastic behavior of structural systems (maps to R&D topic 16)
- Cyber-based tools for improved sharing of information and resources (maps to R&D topic 17)
- Performance-based design for multiple hazards (e.g., wind & wave) (maps to R&D topics 18 and 30)
- Component testing to establish acceptability criteria (maps to R&D topic 26)
- Apply performance-based design framework to evaluate existing codes & standards

D.2. Subject Area 2: Improve Windstorm and Coastal Inundation Design and Community Resiliency

Program Element 3: Support the development of technical resources (e.g., guidelines and manuals) to improve windstorm and coastal inundation engineering practice.

- *Evaluation of resistance of existing buildings (maps to R&D topic 14)*
- *Single wind procedure, including building shapes, attachments, emerging (green) technologies (e.g., solar panels) (maps to R&D topic 8)*
- *Better flood load calculation procedure (maps to R&D topic 23)*
- Low probability of exceedance hazard maps (maps to R&D topics 1 and 22)
- Design for single family houses against tornado (maps to R&D topic 9)
- Setting performance goals for expected wind events (maps to R&D topic 15)
- System-based standards for multi hazard community resilience (maps to R&D topic 18)
- Flood data damage set, damage functions/fragility (maps to R&D topic 26)
- Improved means for risk communication including replacement of deterministic lines on flood maps
- Effective attachment of cladding – includes roof edges

Program Element 4: Make evaluated technology available to practicing professionals in the windstorm and coastal inundation design and construction communities.

- *Wind characteristics & loads for non-conventional boundary layer wind (maps to R&D topic 3)*
- *Integrated portal – GIS risk-based lookup for all-hazards engineering data (maps to R&D topic 17)*
- *Centralized website portal of publications and data (maps to R&D topic 17)*
- Computational Fluid Dynamics – move into practice (maps to R&D topic 7)
- Advanced hybrid wind resistance testing for building components (maps to R&D topic 11)
- Vulnerability assessment – repair or replace, each type of building envelop component (maps to R&D topic 14)
- Benchmark proven models with new techniques (e.g., computational fluid dynamics) to increase the number of tools for practitioners to use
- Cost-benefit improvements for mitigation & code+ construction so more benefits can be considered
- Education/training/certification of design & building professionals

Program Element 5: Develop the technical basis for windstorm and coastal inundation engineering to support community resilience within an all-hazards framework.

- *Coupled modeling (maps to R&D topics 18 and 23)*
- *Resiliency linked to sustainability (maps to R&D topic 18)*
- *Database improvements (maps to R&D topic 17)*
- Laboratory tests (maps to R&D topics 6, 11, and 23)
- Linkage with other professional models – social and economic (maps to R&D topic 18)
- Probabilistic Risk Analysis – need statistics of local storm and surge systems incidence (maps to R&D topic 22)
- Guidance for practitioners and consumers

- Stress tests: HAZUS as a possible point of departure
- Field measurements for model validation (maps to R&D topic 23)

ACRONYMS

AIA:	American Institute of Architects
ASCE:	American Society of Civil Engineers
ASTM:	American Society for Testing and Materials
ATC:	Applied Technology Council
BIM:	Building Information Modeling
BSE:	Base Flood Elevation
C&C:	Components and cladding
CFD:	Computational fluid dynamics
CFR:	Code of Federal Regulations
CSTORM-MS:	The U. S. Army Corp of Engineers Research and Development Center's Coastal Storm Modeling System
FEMA:	Federal Emergency Management Agency
HVAC:	Heating, ventilation, air conditioning
MWFRS:	Main wind force resisting system
NFIP:	National Flood Insurance Program
NOAA:	National Oceanic & Atmospheric Administration
NSF:	National Science Foundation
NSSA:	National Storm Shelter Association
NWIRP:	National Windstorm Impact Reduction Program
NWS:	National Weather Service
USACE:	U. S. Army Corps of Engineers
USGS:	U. S. Geological Survey

REFERENCES

AAWE, (1997). *New Opportunities to Reduce Wind Hazard Losses and Improve Quality of Life in the USA*, American Association for Wind Engineering, 72 pp.

AAWE, (2004). *Wind Engineering Research and Outreach Plan*, AAWE in collaboration with ASCE, American Association for Wind Engineering, http:// www.aawe.org/ docs/ news/ Wind_Eng_Report_May04r2.pdf, last accessed on September 26, 2013.

AAWE, (2010). *Proceeding of International Workshop on Wind Engineering Research and Practice: Current State of the Art and Future Needs/Plans*, American Association for Wind Engineering.

ASCE/SEI, (2010). *Minimum Design Loads for Buildings and Other Structures*, Structural Engineering Institute of the American Society of Civil Engineers, Reston, Virginia.

ATC, (2003). *The Missing Piece: Improving Seismic Design and Construction Practices*, ATC-57 Report, Applied Technology Council, Redwood City, California.

David, J. R., et al, (2010). *Toward the Probabilistic Simulation of Storm Surge and Inundation in a Limited-Resource Environment*, American Meteorological Society, Boston, Massachusetts, 22 pp.

Department of Homeland Security, Science and Technology Directorate, (2008). *Coastal Hazards Colloquium*, University of North Carolina at Chapel Hill, Chapel Hill, North Carolina, http://www.hsdl.org/?view&did=30442, last accessed on September 26, 2013.

FEMA, (2008). *Guidelines for Design of Structures for Vertical Evacuation from Tsunamis*, FEMA P-646, Federal Emergency Management Agency, Washington, D.C.

FEMA, (2011). *Coastal Construction Manual*, FEMA P-55, Federal Emergency Management Agency, Washington, D.C.

FEMA, (2013). *Wind Retrofit Guide for Residential Buildings*, FEMA P-804, Federal Emergency Management Agency, Washington, D.C.

Federal Reserve, 2012, "Housing Starts: Total: New Privately Owned Housing Units Started, Economic Research", St. Louis, Missouri.

Feyen, J. (2010). *NOAA's Storm Surge Roadmap: Transition Research to NWS Operations,* http://www.nws.noaa.gov/mdl/RITT/forum/briefings/SurgeRoadmap_RITT_21July2010.pdf, last accessed on September 26, 2013.

Gopu, V. K. A. (2010). *Wood Engineering Challenges in the New Millennium – Critical Research Needs*, New Orleans, Louisiana.

Irwin, P. (2011). Wind Engineering Research Needs, Technical Council on Wind Engineering, American Society of Civil Engineers, Reston, Virginia, http:// www.aawe.org/ docs/ news/WindEngineeringResearchNeeds%282011%29-07.pdf, last accessed on September 26, 2013.

Jain, A., Srinivasan, M. & Hart, G. C. (2001). "Performance Based Design Extreme Wind Loads on a Tall Building," *The Structural Design of Tall Buildings*, Vol. 10, pp. 9-26, John Wiley & Sons, Ltd., Chichester, West Sussex, England.

Massey, T., Wamsley, T. & Cialone, M. (2011). "Coastal Storm Modeling – System Integration", US Army Corps of Engineers, *Proceedings of the 2011 Solutions to Coastal Disasters Conference, Anchorage*, Alaska, June 26 to June 29, 2011.

Meade, C. & Abbott, M. (2003). *Assessing Federal Research and Development for Hazard Loss Reduction*, RAND Science and Technology Policy Institute for Office of Science and Technology Policy, Arlington, Virginia.

National Academies Press, (2012). *Disaster Resilience: A National Imperative*, The National Academies Press, Washington, D.C.

National Research Council, (1993). *Wind and the Built Environment, U.S. Needs in Wind Engineering and Hazard Mitigation*, Washington, D.C.

National Science Board of the National Science Foundation, (2007). *Hurricane Warning: The Critical Need for a National Hurricane Research Initiative*, Arlington, Virginia, http://www.nsf.gov/nsb/committees/archive/hurricane/ initiative.pdf, last accessed on September 26, 2013.

National Science and Technology Council, Subcommittee on Disaster Reduction, (2005), *Grand Challenges for Disaster Reduction*, http:// www.nehrp.gov/ pdf/ grand challenges.pdf, last accessed on September 26, 2013.

National Science and Technology Council, (2006). *Windstorm Impact Reduction Implementation Plan*, http:// www.sdr.gov/ docs/ Windstorm%20Impact%20 Reduction %20Implementation%20Plan%20FINAL.pdf, last accessed on September 26, 2013.

National Science and Technology Council, Subcommittee on Disaster Reduction, (2008). *Grand Challenges for Disaster Reduction*, http:// www.sdr.gov/ docs/ Grand Challenges Second Printing.pdf last accessed on September 26, 2013.

NIST, (1995). *Proceedings, Workshop on Research Needs in Wind Engineering*, NISTIR 5597 Report, National Institute of Standards and Technology, Gaithersburg, Maryland.

NIST, (2009). *Research Required to Support Full Implementation of Performance- Based Seismic Design*, NIST GCR 09-917-2 Report, National Institute of Standards and Technology, Gaithersburg, Maryland, 81 pp.

NIST, (2013). *Developing Guidelines and Standards for Disaster Resilience of the Built Environment: A Research Needs Assessment*, NIST Technical Note 1795, National Institute of Standards and Technology, Gaithersburg, Maryland.

NOAA, (2013a). *Natural Hazard Statistics*, National Oceanic & Atmospheric Administration, National Weather Service, Office of Climate, Water, and Weather Services, http:// www. nws.noaa.gov/om/hazstats.shtml, last accessed on September 23, 2013.

NOAA, (2013b). *National Coastal Population Report, Population Trends from 1970 to 2020*, National Oceanic and Atmospheric Administration, Washington, D.C.

Obama, B. (2013). Presidential Policy Directive Critical Infrastructure Security and Resilience, U.S. Govt. Print. Off. http://www.whitehouse.gov/the-press- office/ 2013/ 02 /12/ presidential-policy-directive-critical-infrastructure- security-and-resil.

Texas Tech University, (1999). *Windstorm Mitigation Initiative, Retrofit Workshop Report*, Lubbock, Texas.

U.S. Census Bureau, (2011). *Housing and Household Economic Statistics Division, 2011*, http://www.census.gov/hhes/www/housing/census/historic/units.html.

USACE, 2002-2006, *Coastal Engineering Manual*, U. S. Army Corps of Engineers, Vicksburg, Mississippi.

PROJECT PARTICIPANTS

National Institute of Standards and Technology

Eric Letvin
National Institute of Standards and Technology
Engineering Laboratory (MS8611)
100 Bureau Drive
Gaithersburg, Maryland 20899

Marc Levitan
National Institute of Standards and Technology
Engineering Laboratory (MS8611)
100 Bureau Drive
Gaithersburg, Maryland 20899

NEHRP Consultants Joint Venture

Applied Technology Council
201 Redwood Shores Parkway, Suite 240
Redwood City, California 94065
Consortium of Universities for Research in Earthquake Engineering
1301 S. 46th Street, Building 420
Richmond, California 94804

Joint Venture Management Committee

James R. Harris
J.R. Harris & Company
1775 Sherman Street, Suite 1525
Denver, Colorado 80203
Robert Reitherman
Consortium of Universities for Research in
Earthquake Engineering
1301 S. 46th Street, Building 420
Richmond, California 94804
Christopher Rojahn
Applied Technology Council
201 Redwood Shores Parkway, Suite 240
Redwood City, California 94065
Andrew Whittaker
University at Buffalo
Dept. of Civil, Structural, and Environmental
Engineering, 230 Ketter Hall
Buffalo, New York 14260

Windstorm and Coastal Inundation Program Committee

Jon A. Heintz (Program Manager)
Applied Technology Council
201 Redwood Shores Parkway, Suite 240
Redwood City, California 94065
Leighton Cochran
Leighton Cochran Consulting
35 Astor Terrace, Suite 504
Spring Hill, 4000, Queensland Australia
Kurtis R. Gurley
University of Florida
Dept. of Civil and Coastal Engineering
P.O. Box 116580, 365 Weil Hall
Gainesville, Florida 32611-6580
Spencer Rogers
North Carolina Sea Grant

5600 Marvin Moss Lane
Wilmington, North Carolina 28409
Donald R. Scott
PCS Structural Solutions
1250 Pacific Avenue, Suite 710
Tacoma, Washington 98402
Joseph B. Shepard
Haskell
111 Riverside Avenue
Jacksonville, Florida 32202
Thomas Lee Smith
TLSmith Consulting Inc.
16681 Boswell Road
Rockton, Illinois 61072

Project Technical Committee

William L. Coulbourne (Project Tech. Director)
Applied Technology Council
19266 Coastal Highway, Unit 4, #19
Rehoboth Beach, Delaware 19971
Jon Galsworthy
RWDI
650 Woodlawn Road West
Guelph, Ontario, Canada N1K 1B8
Horia Hangan
University of Western Ontario
Western Engineering
Spencer Engineering Building
London, Ontario, Canada N6A 5B8
Christopher Jones, P.E.,
Christopher P. Jones Consulting
5525 Jomali Drive
Durham, North Carolina 27705
Chris Letchford
Renessalear Polytechnic Institute
Department of Civil and Environmental Engineering
110 Eighth Street
Troy, New York 12180
Thomas Lee Smith
TLSmith Consulting Inc.
16681 Boswell Road
Rockton, Illinois 61072

Publication Services

Amber Houchen
Applied Technology Council
201 Redwood Shores Parkway, Suite 240
Redwood City, California 94065

End Notes

[1] As defined by NIST, the term *measurement science* includes "the development of performance metrics, measurement methods, predictive tools, and protocols as well as reference materials, data, and artifacts; the conduct of inter-comparison studies and calibrations; the evaluation and/or assessment of technologies, systems, and practices; and, the development and/or dissemination of technical guidelines and basis for standards, codes, and practices—in many instances via test beds, consortia, and/or other partnerships with the private sector."

[2] Fatality and property loss figures developed from data published by the National Oceanic and Atmospheric Administration (NOAA, 2013a)

[3] NEEShub, the NEES platform for research, collaboration, and education, is located at http://nees.org/

INDEX

#

A

B

C

D

E

F

G

H

I

J

K

L

M

N

S

T

U

V

W

Y